"A timely, practical, and concise guide providing insights to improving one's communication skills. This is an essential read for everyone in the profession of pharmacy (students, faculty, practitioners) who desires to communicate more effectively."
—Alan Hanson, University of Wisconsin School of Pharmacy

"Information Technology offers a global pool of talented professionals or 'geeks.' The definition of 'geek' perhaps is an IT professional who has not had the benefit of this insightful book on how to make the transformation. Seasoned veteran or beginner, the book is a great read!"—Tom Conway, Vice President, IBM Global Technology Services

"As a corporate lawyer at one of the world's leading science and technology hubs, I recommend this book as a must-read for my client base."—James Forrest, Corporate Attorney

"This practical guide is the result of leading thousands of engaging seminars for physicians and scientists across the United States. St James and Barnard have condensed the information, emphasized the best of it, and put in one efficient volume the equivalent of a Master's degree in professional and interpersonal communication."—Paul Casella, MFA, co-author of *Writing, Speaking, and Communication Skills for Health Care Professionals*

"I have found it difficult to find a book that helps students gain the professional and interpersonal skills they need in a scientific discipline. *Listen. Write. Present.* is that book."
—Tim Marks, Campbell University

Listen. Write. Present.

Listen.
Write.
Present.

**The Elements for Communicating
Science and Technology**

Stephanie Roberson Barnard and
Deborah St James

Yale UNIVERSITY PRESS

New Haven and London

Yale University Press books may be purchased in quantity for
educational, business, or promotional use. For information, please
e-mail sales.press@yale.edu (U.S. office) or sales@yaleup.co.uk
(U.K. office).

Set in Adobe Garamond and Calibri types by Keystone Typesetting,
Inc. Printed in the United States of America.

Library of Congress Cataloging-in-Publication Data
Barnard, Stephanie.
Listen. write. present. : the elements for communicating science and
technology / Stephanie Roberson Barnard and Deborah St James.
p. cm.
Includes bibliographical references and index.
ISBN 978-0-300-17627-8 (pbk.)
1. Communication of technical information. 2. Communication in
science. I. St. James, Deborah. II. Title.
T10.5.B36 2012
601'.4—dc23   2011022720

A catalogue record for this book is available from the British Library.

This paper meets the requirements of ANSI/NISO Z39.48-1992
(Permanence of Paper).

10 9 8 7 6 5 4 3 2 1

Contents

Foreword

For scientists, communication should be about getting through to people, helping them understand the importance of the concept, and putting the idea in terms that are scientifically correct but also, and equally important, comprehensible to a variety of audiences.

Communication is one of the most complex processes that happen among human beings. Taking it for granted or not paying enough attention to its intricacies is one of the most frequent mistakes that scientists make. There is a whole field on communication theory devoted to unlock, expose, and improve the way we communicate. From the theories of Ferdinand de Saussure about signs and signifiers to the developments of Umberto Eco in the field of linguistics, evidence and theory reinforce the importance of using the correct tools to communicate our thoughts to other people.

Good leaders must learn to communicate not only within their field of expertise but also to reach people outside their field of authority, influence, and passion. A good scientist should be able to defend a theory in an oral presentation at a key scientific congress as well as put together a strong business case for that same idea in front of a corporate board of directors. Many of us trained in science did not learn these skills in school. *Listen. Write. Present.* is aimed at closing that gap. This book takes the scientist beyond the classroom and beyond the bench into the real world. This book is not about getting a few writing or speaking tips; it is about getting tools that will help you approach, diagnose, and implement a plan to get your message across to all of your intended audiences. Whether you're seeking a promotion, interviewing for a new job, making a case for a bigger budget, managing a difficult employee or patient, or presenting at an international convention, *Listen. Write. Present.* provides the tools to implement an effective communication strategy.

As scientists, we mustn't be lulled into thinking that being a content expert is enough. The way we communicate our message is just as important as the content itself. This small yet wise book reminds us that true leaders transcend their circles of expertise and never forget that for each audience there is a different message and a different way of delivering it.

You already have the potential; this book will unlock it for you.

Rodolfo Chaij, MD
Former Senior Director, Medical Affairs
Talecris Biotherapeutics

Preface

WHY DID WE WRITE THIS BOOK?

Check any online job-hunting website for science, technical, pharmaceutical, biotech, and medical jobs, and among the hundreds of listings, you'll find one common requirement: "excellent communication skills." While this general requisite seems easy to fill, most managers, directors, and vice presidents seek employees whose communication skills go beyond the ability to draft a basic email or design a simple slide. These managers want professionals who can communicate ideas to a variety of audiences in a way that garners support, influences people, and solves problems. Organizations expect these professionals to be able to arrive at the job with both the technical training and the communication skills they need to excel.

"We live in an era of communication, interactions, negotiations, and conflict resolutions. The idea of the scientist in the glass tower happily spending endless days in solitary confinement with his or her test tube, isolated from the environment, is a literary myth. The reality is today's scientists and technology specialists must be able to clearly and persuasively translate their research and proficiencies to audiences not fully immersed in their fields of expertise if they hope to succeed," states Rodolfo Chaij, MD, former Professor of Pharmacology and Cardiology at the University of Buenos Aires and former Senior Director of Medical Affairs at Talecris Biotherapeutics.

Despite excellent education and training, many science and technology professionals lack these essential communication skills, or "soft skills." In the journal *Science,* Dr. Clifford Mintz laments how little training in communication skills students in the life sciences receive: "Unfortunately, few academic programs develop these skills in a systematic way; this failure hinders students from landing industrial jobs."[1]

Likewise, in information technology (IT), many organizations emphasize hard skills, such as programming languages, but it is the neglect of soft skills that leads to project failure. "I think most users would agree that when things go wrong, it's not because the person was technically incompetent," says consulting manager Graeme Simsion. "Far more often, there's been a misunderstanding of requirements from lack of the underlying consulting (communication) skills."[2]

Further, science professionals who want to achieve better outcomes with patients, more funding for research, and career advances or promotions must develop their communication skills. Recognizing this need over the past ten years, more and more medical schools, especially pharmacy schools, have begun to include patient communication skills in their curricula. In 2008, Vanderbilt School of Medicine launched a new curriculum that integrated patient communication skills into its coursework.[3] While these schools have added an important piece to the education of their students, they're still deficient in offering courses in presentation, managerial, and writing skills. Graduates are expected to obtain these skills by trial and error through residency and fellowship training or by combining their MD or PharmD degrees with an MBA. For physicians and pharmacists who wish to enter the pharmaceutical or biotech industry, communication skills are paramount. In an article titled "Transforming Clinicians into Industry Leaders," Melanie Staff-Parsons and Dr. William Pullman summarize the required behavioral competencies for medical directors: "In short, it's all about the ability to influence R&D's scope and direction by balancing technical expertise with interpersonal skills."[4]

HOW ARE WE QUALIFIED TO WRITE THIS BOOK?

We're communication consultants who have worked as editors, trainers, public relations experts, writers, and professional speakers for a combined total of more than thirty years. The professionals described above are the people we've trained in universities, hospitals, research groups, Fortune 500 companies, and medical, nursing, and pharmacy schools around the globe. Many of them have approached us after our seminars on presentation, communication, or writing skills to say, "In all of my education and training, no one has ever taught me this." Our greatest reward is when we return to train a new group and discover that these former trainees have advanced to higher positions within their organizations.

WHAT'S INSIDE?

The pace of today's technology-driven world seems almost frenetic. Few people have time for self-directed learning that does not offer continuing education credit. For this reason, we designed this book to be an easy pickup read. As we planned, organized, wrote, and edited this book, we concluded that the best way to improve communication is to focus on six core skills, which serve as our chapter titles.

Plan. Perhaps the easiest way to improve communication is to slow down and think about what you're trying to say. Before you type, speak, or write: stop. A few minutes of planning can save hours of backtracking, misunderstanding, confusion, and correcting. This chapter will teach you how to choose the right mode of communication, plan for an interaction, save time, manage multiple tasks, and set yourself up for success in all communication. The added benefit: you should be able to carve out time for the important tasks of critical thinking and project evaluation.

Listen. Ours is a noisy world. We get bombarded by phone calls, emails, white noise, and other "chatter." As a result, "getting the message" is harder than ever. Active listening is a learned skill that needs to be practiced and refined, not only to "get" the message the first time, but also to show the other person that you "got" it. In this chapter, we discuss the value of listening skills with specific tips on how to use eye contact, how to listen actively (and really get the message), how to interrupt politely, and how to accept criticism with grace. The benefits of acquiring listening skills: being able to retain information and influence others.

Write. As technology advances, one of the biggest shortfalls is in quality writing. Because we've developed a second language of text shorthand, our writing skills have suffered. Meanwhile, employers still want résumés, reports, proposals, blog posts, and emails that are clear, concise, well organized, thoughtful, compliant, and error free. Since you may not have time to take a business writing class, we've pulled together a chapter that teaches advanced writing techniques to help you become a clear, concise, and persuasive writer. The best lessons here: our years of writing and editing experience distilled into one easy-to-read chapter.

Present. Interpersonal communication and presentation skills are paramount to career success. After all, the best idea in the world is worthless if you can't explain it to someone else. Today clear, well-organized presentations are equated with clear, well-organized thinking. This chapter is filled with indispensable advice on telephone, webinar, and teleconference etiquette, and presentation and interpersonal communication skills. We've coached thousands of scientists and professionals on how to be better presenters and we've packed our best tips in here.

Meet. A fundamental part of interpersonal communication is meeting. We meet to discuss ideas in a formal gathering. We meet potential and new colleagues and clients. We meet the press when our ideas go public. How can you manage all of these meetings with professionalism and ease? This chapter provides the answers: how to present at a meeting, how to "meet and greet" new business contacts, and how to prepare for a television or radio interview. The greatest tips here are on how to run effective meetings so you can get things done.

Serve. Being able to communicate effectively with customers, colleagues, and superiors is the key to moving up the corporate ladder. If you see yourself as serving others, your communications will take on a whole new meaning and enhance your influence in the workplace and beyond. This chapter focuses on the timeless topics of managing others, collaborating, and dealing with difficult people. These are the communication lessons we learned by trial and error, the ones we wish someone had shared with us when we first embarked on our own careers.

WHY READ THIS BOOK?

Many people have written books on communication skills that serve as great academic pieces and offer interesting theories on why effective communication works.

Our book is different. During our years of training health care professionals, we've learned that most people have already come across the theories behind good communication in a book or class. Our audiences said they needed everyday tips, often referring to them as "the pearls" of good communication. We responded and created workshops to meet this demand. Now we're offering our best, tried-and-true tips for you. Our goal here isn't to discuss theory, but to teach you specific ways to adapt your communication style so that you're more useful in your job (and your life) immediately.

As we wrote this book, we debated how to make it most effective. After all, we could easily have put in a thousand tips or shared hundreds of personal stories and well-researched examples to support our points. We chose not to include these items because we felt that you, our reader, didn't want to wade through extra stuff to get the point.

At first glance you may think the numbered tips and lists make this book too simplistic. In fact, we deliberately chose this style to make the book a user-friendly guide, packed with important details that you can read, recall, and implement easily. The idea is for you to pick up the book any time to read just one section or use as a quick reference.

If possible, we recommend that you read through the entire book once. We know you probably think you don't have time to read it cover to cover. Please try to make time—on your next flight, during your lunch break, or while you wait for an appointment. Why? Each section has different tips and ideas and all of the sections work together. They are deliberately brief and focused, and well worth the few extra minutes.

ACKNOWLEDGMENTS

From the moment we first pitched this book idea to our editor, Jean Thomson Black, she graciously offered advice, encouragement, and guidance to us. Thanks to her and everyone at Yale University Press for high standards of excellence in publishing: Debra Bozzi, Jaya Chatterjee, Wendy DeNardis, Sara Hoover, Linda Klein, Larry Laconi, Ivan Lett, Margaret Otzel, Nancy Ovedovitz, John Palmer, Karen Stickler, Terry Toland, Tanya Weideking, and Jenya Weinreb.

Thanks to Susan Cantrell, Rodolfo Chaij, Susan Goodin, Al Hanson, Tom Holmes, Mark Kritzman, Tim Marks, Jane Sillman, Tom Stocky, Lucas Turton, and Michele Vivirito for offering great advice on how to make the manuscript better.

To freelance editor Kim Hastings, your work is outstanding. You were able to see the trees at times when we were lost in the forest.

One of the challenges of writing with brevity and density is attending to the details. To our summer interns, Kelly Merrick and Bob Tennant: great proofing! To Yale's proofreader, Patty O'Connell, and indexer, Nancy Wolff: excellent work!

Many thanks to Rob Monath for wise counsel.

We are grateful for the many expressions of support from family, friends, and colleagues. Your words of encouragement and pre-orders made us smile!

We dedicate this book to our families. Thanks, David, Callie, Sarah, and Patrick! We love you.

Finally, we must also acknowledge the many teachers who paved the way by teaching us, and the thousands of people who attended our seminars and shared feedback, stories, and ideas. We couldn't have written this book without any of you. We take our own advice and we LISTEN. It's the first lesson for any good communicator.

1 Plan

Let's begin with a story. Mark (not his real name) was a highly successful sales representative in a major biotech company. This "Rookie of the Year" had a PhD in biochemistry, dynamic people skills, a great drug portfolio, and a ripe territory. His sales figures were smashing, and he won the highest accolade in pharmaceutical sales, The President's Club award, where he and his peers were honored at a fancy dinner with all of the company executives. Mark was delighted to share his talents with the leadership team and, after several drinks, approached the company president to explain why he should be promoted. Unfortunately, this tactic backfired, as the president and his peers weren't impressed by Mark's bragging. Mark ultimately left the company to seek the opportunities he lost when he failed to think through this important communication. This story isn't an isolated incident. In our seminars, we often hear about successful people who don't plan important communications and end up with irreparable damage.

The previous example is filled with bad choices: drinking excessively in a professional setting, choosing an inappropriate time and place to discuss career objectives, skipping the chain of command, and dominating the conversation. The biggest problem, however, is that Mark didn't *plan*. In a sales rep's everyday business, speaking with minimal preparation is acceptable. As Mark approached his next career goal (and the company president), both the stakes and the skills needed became higher. To move to the next level, Mark needed to show business acumen through his ability to communicate ideas in an appropriate setting; to collaborate with others rather than showcase individual talents; and to negotiate for a better position by understanding timing, audience, and company protocol.

As a science or technology professional, you understand that part of completing an important task such as learning new material or wading through complex information is setting aside time for critical thinking. The best way to do this is through planning. In this chapter we offer just sixteen tips. That's because we *planned* it carefully. We spent a good deal of time thinking about the most important information to include, researched, and distilled it into the most relevant tips and examples.

PLAN FOR SUCCESS

Planning is a crucial part of effective communication. Accomplished businesspeople will tell you that a strategic business plan is the key to success. Planning helps them wade through ideas, pinpoint the best actions to take, stay focused as the business environment changes, and acquire funding for various projects. All things that you, as a scientist, engineer, medical professional, or computer specialist, must also do.

When you plan before you communicate, you set yourself up for success as a collaborator and negotiator by

- Thinking through what you're going to communicate instead of spewing out irrelevant details
- Making sure you have included all of your ideas instead of just those that pop into your head during the interaction
- Preparing to respond to questions about your idea so you sound articulate
- Refining your ideas through critical thinking
- Saving time in every communication situation including meetings, telephone conversations, presentations, email writing, and supervising others

Whether you're involved in launching a new product, organizing a wedding, or just running to the supermarket, you need a plan. A study from the University of Florida found that supermarket shoppers without a precise list took longer to shop and spent more money on their items than those who planned ahead.[1] At work, as with shopping, planning will save you both time and money.

Are there times in your work life when you feel so overwhelmed by meetings, deadlines, and projects that you simply don't have time to think? For many people, diving into the tasks of the day seems like the most logical way to manage an overwhelming "to do" list. Yet these task-oriented people, and their procrastinating counterparts, miss a great opportunity to sharpen their projects by taking time to plan and think about them before they begin. You may have heard some of the concepts in this section before. As you read these planning tips, think of how to implement them immediately as both time-saving and negotiation tools for everyday situations.

1. Start your day with a list of tasks. Take ten minutes each day to prioritize your tasks. (You may want to do this at the end of the business day so you can start the next day with the list in hand.) Then decide which ones require planned agendas.

2. Choose the most pressing tasks first. This will enable you to give your most difficult and daunting projects your best creativity and motivation.

3. Stop trying to multitask. Communicating requires you to focus. When you multitask, you give each task only part of your attention, which invariably makes it take longer to complete.[2] For example, trying to talk on the phone while you read your email seems like a good idea until you have to ask the other person to repeat what he or she said. It's kind of like trying to cook two things that require stirring at once: something's bound to get scorched.

4. Create an agenda for all communication. Never pick up a phone, start typing an email, start a voicemail, initiate a negotiation, or go into a meeting without a clear idea of what you plan to say. You have a responsibility not to waste other people's time just as they have a responsibility to you. Creating an agenda not only saves time; referring to it keeps you from stumbling about as you try to speak. Try jotting down a few notes on a pad of paper—a list or mind map. The process of writing ideas down will not only help you retain them but also think through them before you communicate.

5. Make appointments with specific start and stop times (and stick to them). When you make appointments with specific start and stop times, you define how long you're going to spend on the topic or task. Everyone involved in the meeting knows your plan, which expedites the process and encourages collaboration.

By the way It's okay to excuse yourself from a meeting or phone call. If you have an impromptu meeting or phone call, or a planned meeting or phone call that is going past the designated stop time, tell the other person when you need to go. For example, "Sam, I'm so glad you stopped by . . . I want to discuss this more but I have a conference call at 3:00."

6. Find out the best way to communicate with others. Take a moment to think about how the listener best receives information: person-to-person, email, telephone, early in the day, over lunch? If someone always arrives early to work, make an effort to get to him or her early.

7. Avoid giving bad news or having difficult conversations through email, voicemail, or text messaging. Always have these challenging conversations in person.

SATISFY YOUR NEED TO BE SOCIAL

Assuming people can meet basic physiological and safety needs, most psychologists agree that the next step is social satisfaction. Many people find this satisfaction (and the higher-level needs of self-esteem, self-actualization, and transcendence) in their work relationships.[3] Thus, cultivating the social part of your work relationships can take you to the next level. Remember, people do business with people they like and trust. The sales rep in the story at the beginning of this chapter may have been pulling in the business, but he wasn't liked or trusted by his peers. Recall a situation when you and another person collaborated on a successful project or negotiated a great deal. What made this inter-action work? Most likely you and the other person liked and trusted each other. The following tips offer ways to cultivate such relationships.

8. Take time to get to know your clients, colleagues, and co-workers. Establish rapport and cultivate a collaborative rela-tionship by finding out about others' interests (check out the pictures in their offices for clues) and inquiring about them. If you have never been to their offices, look them up on Google or their company's website. Always keep your personal conversa-tions light and professional.

9. Identify the person who can get what you need. Take the initiative to find out who is the "face of the company" and make contact with this person. Many people think that big shots in big organizations are unavailable. You'll be surprised who will speak to you if you simply ask. One trick: try reaching them before or after regular office hours.

By the way Don't overlook a person's assistant as a resource. Often you can open locked doors if you have a relationship with an assistant.

10. Adopt a No News = No Action policy. To unclutter your in-box, suggest to others that no news means the plan stays the same. For example, in a voicemail or email, say, "If I don't hear from you, I'll assume we don't need to meet this week." This kind of communication can save you and your colleagues lots of time exchanging redundant information.

11. Schedule meetings at the other person's office. Meeting at the other person's office may seem as if you're giving away control of the meeting. In fact, when you meet at the other person's office, you give the other person the feeling of power while you maintain control of your schedule. When you need to stop the meeting, instead of asking the other person to leave, you get to leave.

12. Keep your colleagues informed. You can do this with a simple email, phone call, or passing comment, for example, "I just wanted to let you know I'm still waiting to hear from Dr. Ahmed on that grant." This gesture shows respect for the other person's investment in your project.

13. Sharpen your skill set. You have already started the most important step for improvement: reading this book. Another great way to be a better communicator is practice. Take on that extra writing project so you can learn to write more effectively. Attend a workshop or take an online class to learn a new skill. Ask to run a meeting so you can sharpen your facilitation skills. Seek out public speaking opportunities so you can work through your stage fright. We advise our audiences to "take baby steps." Choose a nonthreatening environment (possibly away from work) to write a proposal, deliver a presentation, or run a meeting. A great place to get free training is to volunteer for a local charity. We have made many new business contacts over the years while volunteering.

14. Think before you speak. Pause for a few seconds before you speak. Use this bit of time to make sure what you want to say makes sense and adds value to the conversation. You may prevent yourself from interrupting or saying something that you'll regret later.

15. Respond within twenty-four hours. If at all possible, reply to emails, voicemails, and other messages promptly. If you're overwhelmed with daily messages, send a brief reply of "Thanks for your message. I'll be able to reply in detail by (date)," or "I'm swamped with a financial proposal . . . I have your message and I want to hear your opinions when I can give you my full attention. I'll plan to contact you by next Tuesday to discuss."

16. Slow down. Think about a time when you accidentally overlooked or forgot something. Did you happen to be rushing? Over the years, we've heard many confessions from our seminar participants on how they rushed through various communications with disastrous results. One mishap that recurs in all cultural and academic contexts: accidentally hitting "reply all" on an email that should have gone to only one person.

2 Listen

Stop, look, listen. Where have you heard these instructions before? It was when you took driver's education and learned about railroad tracks. The long stretches of metal tracks seem harmless until a speedy locomotive zooms by on them. There's a reason this slogan sticks: it's sage advice packed into a nifty three-verb phrase. Stop, look, listen. You can avoid a catastrophic collision with a train if you heed this advice.

Listening seems like such a simple task that everyone must be good at it. On the contrary, listening requires more than ears; you have to use your brain. Think about the people you know who are highly successful in their fields: the doctor who seems to spend lots of time with patients; the colleague who consistently nails what needs to be done; the project manager who negotiates consensus with difficult clients. What separates these people from others? They are skillful listeners. You can cultivate your listening skills, too. Just learn how to be an active listener.

LISTEN ACTIVELY

Eye contact is the most powerful nonverbal communication tool you own—use it wisely. A friend once told us of her dissatisfaction with her doctor's lack of eye contact. "He just types on the computer the whole time," she said. "I think I might change to a doctor who can look me in the eye." Her grievance is nothing new. Patients, customers, and colleagues in all professions complain when others forgo eye contact. In human interactions, we associate eye contact with honesty and empathy. Multiple scientific studies show there is a correlation between eye contact and improved social interactions. In fact, direct eye contact between people involved in an interaction leads each to perceive the other more positively. By contrast, lack of eye contact triggers a negative response in the brain.[1] In defense of the doctor, we might acknowledge that time is limited, help may not be available, and liability always looms. Typing-while-talking is a great way to document the details. So how can the doctor improve eye contact and still capture the information?

17. Make a point to use eye contact. Looking people in the eyes shows confidence, trust, respect, and empathy. Empathy and respectfulness are key components of effective communication. The doctor in the above example should make a point to look up and make eye contact throughout the exchange. He might try looking up when he makes a comment, or when the patient says something new or significant. This doctor (and anyone else who needs to capture information while communicating) can try one of these techniques:

▪ Asking, "Is it okay for me to type while we talk?" If the other person says "no," then explain why you need to capture this important information.

- Saying, "I need to make sure I get this down. Please pardon me for a moment while I type."
- Training an assistant to take notes during the interaction.

18. Rotate where you look on the other person's face. First look at the forehead, then the mouth, then the eyes, and so forth. The other person will perceive that you're making eye contact while you get the benefit of a visual break.

19. Be aware of cultural differences in eye contact. Eye contact is a powerful tool and some cultures find sustained eye contact offensive. Watch for cues from other people and, if necessary, adapt to the other person's style. It's okay to look away, look down at your notes, or look at a visual aid for a moment.

20. Don't send a text message, check email, or perform unrelated tasks during a conversation. We realize that today's technology sometimes gives people the excuse to "check the handheld" during an interaction. If you must, be sure to say, "excuse me a moment," and swiftly return your attention to the other person.

21. Acknowledge what the other person has said with verbal cues ("okay," "yes," "I see").

22. Don't interrupt others or finish their sentences. Not interrupting others, especially those who are having trouble finding the right word, takes great discipline. Our trick: pause and count to three before you speak.

23. Echo what others have said. Listen for the message—don't just repeat the words. People don't want just to be heard; they want to be understood. You can echo what the other person said without sounding trite: "Let me make sure I understand . . ."

Our advice is to step out of the conversation momentarily to observe the other person(s) in a freeze frame. Notice what is really happening as the conversation progresses. When negotiating, use echoing to show the other person what he or she said, and how it relates to your argument.[2] For example, "So you're saying although we have the money in the budget to purchase the new, more accurate oscilloscopes, you'd prefer to continue to use our current models."

24. Smile, nod, and acknowledge the speaker—and mean it. Really focus on what the person is saying and not just on the words. Truly effective communication requires your full attention. It's better to spend a few minutes concentrating on the other person's message during a conversation than wasting time trying to remember what he or she said because you were trying to do something else. It's okay to write or type notes as long as you ask permission first.

By the way The advantage of active listening is that it takes less time to focus and hear the person once than to half listen and miss the point.

25. Develop situational awareness skills. Part of being a good communicator is being able to observe your surroundings and adapt to them. Are you talking too loudly? Is the person sitting at the next table in the cafeteria eavesdropping on your confidential conversation? Is the person standing nearby waiting to speak to you? An easy way to check your surroundings is to stop talking momentarily, look, and listen.

26. If you must interrupt, do so by saying the other person's name. To keep the conversation moving, sometimes you'll need to interject a comment when someone else is speaking. The best way to show respect and get the speaker's attention is to say his or her name first. For example, "Charlie, let me interrupt for a minute . . ."

By the way If you accidentally speak too soon, simply say, "I'm sorry, please go on."

27. Learn to cultivate relationships. Most people prefer to collaborate with others who share their interests. You can do this easily by cultivating relationships. Notice how the following tips all rely on excellent listening skills.

- **Seek common ground.** Ask questions so you can learn about your colleagues.
- **Learn names.** This may take a bit of extra effort: try making up a mnemonic, repeating the name (at least three times) in your conversation with the person, or writing it down.
- **Note important information.** Take a moment to jot down personal details about a client or colleague shortly after you first meet. Review them before you meet again. Nothing impresses others more than recalling their children's names, the way they take their coffee, or the location of their last vacation. If it was important enough for them to mention, it's probably important enough to note for later.

SKILL BUILDER HOW TO SAY LESS AND EXPRESS MORE

It's very tempting to talk unnecessarily. After all, humans thrive on contact with others. Yet there are some great reasons to be quiet.

| Unnecessary comments can
| Take up time without adding value.

For example:

DAMIEN: "Can you meet next Thursday?"
MEGAN: "No. I have a doctor's appointment in the afternoon. You know, because I'm going to have surgery for my shoulder. I need to find out how long I'll need to recover in case I need to arrange child care . . ."

A more efficient way for Megan to respond is
"I can do Thursday morning or anytime Friday. What do you prefer?"

Commit you to do things that you might not want to do.

For example:

SARAH: "We really need to get started on that budget proposal. I think maybe we should add a section on new employees."
BOB: "Sounds great." *Here's where Bob should stop talking.* "After I finish my section, I'll work on that." *Bob just gave himself another assignment.*

Shed light on what you don't know. Especially when you're new to a job or assignment, it's best to err on the side of quiet.

Prevent you from getting the message. When you're talking, you're not listening and therefore not getting the message. You don't have to tell everything you know to appear smart. In fact, many skilled communicators use nonverbal gestures (eye contact, smiling, nodding) and short affirmation comments ("okay," "I see," "yes") to show understanding.

KEEP AN OPEN MIND

28. Listen for opportunities when volunteering is a good idea. There are many occasions when you'll want to volunteer to do an assignment: when your boss suggests it, when it falls within your job description, or when you know it will help advance your career.

29. Know it's okay to say "no." When someone asks you to do something that you don't want to do, simply say "no." If you toss out an excuse, you give the other person an opportunity to come back with a solution.

For example:

> LUCAS: "Ana, do you think you might be able to help us with the charity auction Thursday night?"
> ANA: "No. I have a meeting that night."
> LUCAS: "That's okay. You can just come after your meeting."

30. Sit down for the conversation; stand up when you're done. Assuming that the other person is already seated, you look more relaxed and open to ideas when you sit down. You also place yourself at eye level with the other person, which gives you the advantage of appearing gracious. (This trick is especially helpful if you're significantly taller or shorter than the other person.) When you stand up, you signal that you need to leave.

31. Take notes religiously. Have separate notebooks, sections of a notebook, or computer files for different projects. This way you can pick right up where you left off the next time you meet. It also helps you stay focused and organized, meet deadlines, and show professional competence.

32. Summarize with paraphrases and transitions. One of the easiest ways to accomplish this task is the Wrap and Roll Technique. First you *wrap* or summarize the conversation: "I understand that you want to get this project done by Tuesday." Then you *roll* or transition with a departure statement: "I guess I need to get moving on my part." Stand up or move to signal your departure.

33. Be aware of your tone. People listen to the tone of your voice more than the words you speak. Watch out for sarcasm and inappropriate humor in your voice. Some people will take offense to these. Even the most benign statements can be misunderstood if you say them sarcastically.

34. Make sure your nonverbal gestures match what you say. Your body position also speaks volumes. People interpret lack of eye contact, crossed arms, and engaging in other activities (such as texting) as negative.

35. Ask for feedback and change accordingly. Remember, true leaders want constructive criticism. They know that every person on the team has something to contribute. To encourage collaboration, they seek ideas from all levels—from the CEO right down to the administrative assistants. Some of the best ideas come from the most unlikely people.

36. Accept criticism with grace. Particularly when you begin a new job or project, you feel a sense of ownership that makes you protect and defend your ideas. This display of conscientiousness is a sterling quality, unless it continually lands you in a defensive posture. Despite your qualifications, you may not be the one with the best answers. That's okay. Take what others suggest and assimilate it with what you know for an even better idea.

37. Seek knowledge. You have to remain current on technology, trends, and company policies to keep from appearing stagnant. Take a class, learn a new computer program, visit other departments, and do whatever is necessary to keep learning. Most people enjoy teaching others. Who knows, the manager from the other department who explains a concept to you over lunch may someday be your boss.

38. Build alliances. Get to know people in other departments and learn their names—not just the "bigwigs" but people at all levels. You'll be surprised how your personal contacts can become career assets later. At the very least, having a relationship with fellow employees such as the custodian will ensure you can get into your office without a key.

3 Write

Today we're overwhelmed with written communication. Our main goal is to get rid of it. Deleting brings us a sense of accomplishment. As a result, if you want to get your written message across, you must be clear, concise, and well organized. Written communication is everywhere: text messages, emails, blogs, tweets, and, of course, letters, proposals, and reports. Despite all of this written communication (or perhaps because of it), many people struggle to pull together thoughtful documents. Writing helps you gather, sort, and express ideas. Through this process, writing also helps you develop critical thinking skills that enable you to solve problems.

In a study conducted by the American Society for Engineering Education, researchers analyzed thirty-eight skills needed to succeed in the industry.[1] Technical writing ranked as number two. Several other communication skills ranked in the top nine:

- Public speaking (fourth place)
- Working with individuals (sixth place)

- Working with groups (seventh place)
- Talking with people (ninth place)

Of all the tips in this book, the ones in this chapter on writing have the greatest potential to help you further your career. To acquire funding for your research or a loan for your business, you must be able to write a scientific paper or business proposal. To forecast sales and document productivity, you must be able to write a strategic plan. To change jobs, negotiate a contract, or ask for a raise, you must be able to write letters and reports. To persuade others, negotiate a deal, or collaborate on a project, you must be able to write clearly, concisely, and confidently. Ultimately, to achieve a leadership position in any organization, you must be able to write effectively. Here are some of our best tips for improving your writing.

39. Read more. Ask any published author or successful editor an easy way to improve your writing and you'll invariably hear: read. Not only will you gain the obvious benefits of greater knowledge and increased vocabulary, you'll also see how other writers craft sentences and argue points to make them stronger. We suggest reading the following.

General documents: newspapers, magazines, websites, blogs, and general nonfiction. Reading these sources will expose you to a variety of topics to learn trends in the marketplace; local, regional, and global news; and concepts in pop culture and human nature. You'll benefit from hearing how laypeople describe your area of expertise and by being able to place your work into the context of everyday life in our global economy.
Specific to your trade: scholarly journals, websites, blogs, and academic books. Reading these sources will help you expand

and refine your ideas to become a critical thinker, thought leader, and collaborator. Although you may be a specialist in your field, you're not alone in your expertise. Seeking knowledge from others in your trade will enhance your arguments. **Specific to your audience:** websites and marketing materials written for the lay audience about your field. Reading these sources will help you understand the best way to communicate to your audience. Even if you're immersed in research, technology, or academics, you still need to grasp what consumers understand about your work, as everything eventually has market-driven implications.

Science writing award winners: these will serve as excellent examples of accomplished science and technical writers. Consider the Taylor/Blakeslee and Haseltine Fellowship, the American Institute of Physics Science Writing Award, the Council for the Advancement of Science Writing Award, the Pulitzer Prize for Explanatory Reporting, and particularly the American Association for the Advancement of Science Kavli Science Journalism Awards.

40. Before you ever put pen to paper or hand to keyboard, ask yourself:

- Who is my reader?
- What does he or she need to know?
- What is my purpose?
- What is the goal I hope to accomplish?

41. Consider your reader—less is always better. Even with an excellent education and work experience, most people prefer to read the short version of any story. That is why this book was written with pertinent examples only. To accomplish this style in your writing, make sure you understand the purpose of the document. Ask yourself: *What action do I want the reader to take?*

42. Make written notes before you start typing. When you make notes or outline before you start writing, you force yourself to think about your topic. Even if you only take a few minutes at a break in the day, jotting down your ideas will help you formulate them in your mind, organize them, and eventually get them on paper. Three other great reasons to outline:

■ If you're interrupted while writing, you'll know where to pick up later.

■ You can jot down ideas any time, even if the document isn't open on your computer.

■ You'll be able to think critically about your idea, even when you're not working on it, and possibly generate solutions.

43. Choose a reputable style guide for your trade or industry and use it consistently. A style guide is a worthy investment that will answer questions on grammar, punctuation, and word usage and help you appear polished, professional, and well educated. See the recommended resources at the end of this book or check with colleagues for the one best suited for you.

44. Review samples of good writing. What does the reader expect? Ask your boss or colleagues for samples of acceptable reports, proposals, and letters. Note the strengths of these documents: brevity, active voice, specifics. Emulate what works and use the rest of this chapter to learn other writing techniques.

MISUSED WORDS

Good scientific, technical, or medical writing is directly related to correct word usage, punctuation, and grammar. Errors in any of these realms can cause confusion in understanding—not to mention reflect poorly on you. The following sentences point out some of the most frequent problems we see in the documents we review. This is by no means a complete list. We urge you to double-check a good reference when in doubt. (We have recommended several great style manuals and websites at the back of this book.)

The best way to avoid using the wrong word, which Spell Check often doesn't pick up, is to be alert to the most frequent offenders, and always proofread your documents—even brief emails.

IT'S AND ITS

RULE Use *it's* when you want to say "it is" or "it has." Think of the apostrophe as standing for the *i* in *is* or the *ha* in *has*.
For example:
It's considered an orphan disease. **Or** *It is* considered an orphan disease.
It's been a long time coming. **Or** *It has* been a long time coming.
Use *its* everywhere else.
For example:
The committee will present *its* report at the annual board meeting. (You wouldn't say, The committee will present *it is* report.)
The dog was examined and *its* temperature was 101 degrees.

ENSURE, ASSURE, INSURE

RULE ensure = make certain
For example:
> The customer service rep said, "I will *ensure* you get a full refund."

RULE assure = instill confidence
For example:
> We want to *assure* our customers that the ski lift is safe.

RULE insure = protect against loss
For example:
> You should *insure* the ring before you leave the store.

I.E. OR E.G.

RULE i.e. = Latin for *id est*. It simply means "that is." Use *i.e.* when you want to clarify a point or when you want to say, "in other words."
For example:
> While working in Venezuela, Dr. Dennis became very familiar with mycetoma, *i.e.*, a tumor produced from fungi.

RULE e.g. = Latin for *exampli gratia*. It means "for example." Use it before giving specific examples that support your statement.
For example:
> I will be attending all the major pulmonary meetings this year, *e.g.*, ACCP, ERS, ATS.

Please note:
> When using *e.g.*, don't use *etc.* at the end of the sentence. It's like saying, I'm not going to bother listing them.

ME, MYSELF, OR I

RULE Use *myself* in two situations only:
• To show that you are doing something to yourself.
For example:
I'm going to write *myself* a note so I don't forget to bring in the computer.
• To emphasize what you thought or did.
For example:
I *myself* would never have approved that grant.
Incorrect: Dr. Dixon, Dr. Heron, and *myself* will be welcoming Dr. Amad to our Westside office on Wednesday.
Correct: Dr. Dixon, Dr. Heron, and *I* will be welcoming Dr. Amad to our Westside office on Wednesday.

Incorrect: It was nice of Juan to thank Sally and *myself* for our work on the project.
Correct: It was nice of Juan to thank Sally and *me* for our work on the project.

RULE We often see and hear confusion when it comes to *I* or *me,* particularly when used with other names. The solution is simple. Try removing the other name(s).
For example:
Incorrect: The CEO mentioned Henry and *I* in his presentation about the Alpha project.
Correct: The CEO mentioned Henry and *me* in his presentation about the Alpha project.

RULE Use *me,* not *I* or *myself,* after *at, between, by,* and *to.*
For example:
Wow! That question the Federal Trade Commission directed *at* Henry and *me* (not *I* or *myself*) was tough.

The case report was written *by* Jamaala and *me* (not *I* or *myself*).
Between you and *me* (not *I* or *myself*), that article should have
been rejected.

AFFECT OR EFFECT

RULE affect (v) = to influence
For example:
  The drug can *affect* your appetite.

RULE affect (n) = behavior or outward appearance
For example:
  Classmates who knew the shooter say he had a strange *affect*.
  He was often seen laughing at inappropriate times.

RULE effect (v) = to bring about
For example:
  The newer antipsychotic drugs can *effect* dramatic results in a
  relatively short time.

RULE effect (n) = outcome
For example:
  What is the *effect* of the recent cutbacks on morale?

FEWER OR LESS

RULE Use *fewer* for things you can count; use *less* for things you
can't count.
For example:
  Jeff has *fewer* publications than Susan, but both are *less* qualified than Tariq.

BETWEEN OR AMONG

RULE When there are more than two things or two people involved, use *among;* otherwise, use *between.*
For example:
The Food and Drug Administration (FDA) found the contaminated vial *among* the thirty recovered samples.
Between Dr. Habid and Dr. Cruz, I think Dr. Cruz has the better research skills.

IN REGARDS TO

RULE Keep the word *regard* singular. There is no need for the final -*s.*
For example:
Incorrect: We need to review the data *in regards to* the pathogen safety issues.
Correct: We need to review the data *in regard to* the pathogen safety issues. **Or** We need to review the data *regarding* the pathogen safety issues. **Or** We need to review the data *as it relates to* the pathogen safety issues.

PUNCTUATION

COMMA

Putting the last comma in a series before the word *and* is a style question. Some style guides say put it in (*The American Medical Association Manual of Style* and *The Chicago Manual of Style*) while others say take it out (*The Associated Press Stylebook* and *The Washington Post Deskbook on Style*). Whichever you choose, stick with it. Consistency is important.

RULE Don't use a comma before *and, but, for, so,* or *yet* between an independent clause (a clause that has a subject and a verb) and a dependent clause (a partial sentence).
For example:
  John will probably get the bigger office *but* not get the raise he thinks he deserves.

Do use a comma before *and, but, for, so,* or *yet* when these words are used between two independent clauses (two full sentences).
For example:
  John will probably get the bigger office, *but* I doubt he'll get the raise he thinks he deserves.

RULE Put a comma after an introductory clause. Frequently introductory clauses start with *after, although, because, before, despite, if, since,* or *when.*
For example:
  Although she has an almost perfect GPA, her communication skills are so poor that she'll probably have a hard time in the job market.

RULE Do not use a comma when the clause comes at the end of the sentence.
For example:
  She's going to have a tough time in the job market *despite* her almost perfect GPA.

RULE Put a comma before an introductory phrase with a verb.
For example:
  As you predicted, Julia was an outstanding summer intern.

RULE When a month, day, and year are all used together, put a comma after the day.
For example:
> My date of hire is July 10, 2010.

RULE If one of the three elements (month, day, or year) is missing, you can omit the comma.
For example:
> The final report came in October 2009.
> His supervisor will make a decision by December 15.

SEMICOLON

RULE When the word *consequently, however, moreover,* or *therefore* is used between two independent clauses, put a semicolon before and a comma after.
For example:
> The FDA is coming to inspect the plant; *therefore,* John has asked everyone to postpone vacation plans for that week.

RULE When *consequently, however, moreover,* or *therefore* is used as an interjection, put a comma before and after.
For example:
> Dr. McMahon was the first at the scene. She was, *however,* never questioned by the police.

RULE Use a semicolon between two independent clauses (simple sentences) that aren't joined by a connecting word such as *and, but, for,* or *so.*
For example:
> The CEO resigned on Monday; the board will begin their search for a replacement next week.

RULE Use a semicolon to separate a series of phrases that already have commas.

For example:

Dr. Patel has been invited to give her talk on Equine Embryo Transplants in Raleigh, NC; Davis, CA; Denver, CO; and Lexington, KY.

COLON

RULE If a sentence contains a long list, you can help the reader by introducing it with a colon; however, you must have a complete sentence (subject and verb) before the colon.

For example:

Lupus presents with a wide range of symptoms: swollen joints, fever, fatigue, hair loss, a butterfly-like rash.

RULE Use a colon between two independent clauses when the second independent clause explains or summarizes the first independent clause.

For example:

Strategic Publication Planning is exactly what it implies: we carefully prepare publications with the right message, for the right audience, at the right time.

HYPHEN

RULE Authorities generally agree that the use of hyphens defies hard and fast rules. The best advice is to consult a good dictionary or style guide. One rule that is useful to remember: Hyphens are *not used* after adverbs ending in *-ly*.

For example:
> He is a highly motivated technician.
> The Willis Company is strongly committed to its clients.

QUOTATION MARKS WITH OTHER PUNCTUATION

RULE Periods and commas always come inside the quotation marks. There are no exceptions in American English—whether the whole sentence is quoted or just a word or two.
For example:
> She said our technique was "flawless."
> She said, "Your technique is flawless."
> It was marked "Toxic," but they packed it with the other items anyway.

RULE Question marks go inside the quotation marks in every case except one: when the sentence begins with a question but ends with a quotation that is not a question.
For example:
> I can't believe Marco would ask, "Will there be lay-offs if the Phase II trial fails?"
> Did Marco really tell his group, "If the Phase II trial fails, there will be lay-offs"?

RULE Semicolons and colons always go outside the quotation marks.
For example:
> David said, "I think we should hire the candidate we interviewed last week"; however, by the time we called her, she had already accepted another job offer.
> Be careful with the vials marked "Contaminated": vial 12, vial 34, and vial 67.

RULE Exclamation points always go inside the quotation marks.

For example:

I about jumped off my chair when the lab tech yelled, "Snake's loose!"

GRAMMAR

MISPLACED MODIFIERS

Misplaced or dangling modifiers are phrases that make it unclear who or what is being described.

RULE Place the modifying phrase as close as possible to the word it describes. This usually prevents the problem.

For example:

Incorrect: Is there a doctor with a cell phone by the name of Roberson in the room?

Correct: Is there a Dr. Roberson in the room who is missing a cell phone?

Incorrect: After buying the software, the computer ran more efficiently.

Correct: The computer ran more efficiently after I installed the new software.

Incorrect: Having reread the results, the discussion section finally came together.

Correct: After we reread all the results, we were better able to write the discussion section.

Incorrect: After five ECTs, Dr. Hu was discouraged with the patient's response.

Correct: Dr. Hu is discouraged since the patient has shown little response following five ECTs.

SUBJECT-VERB AGREEMENT

RULE Words between a subject and its verb don't affect its agreement.
For example:
Supervisors at the new facility *say* morale has improved.

RULE When subjects are joined by *either . . . or, neither . . . nor,* or *not only . . . but (also),* the verb must agree with the subject closest to it.
For example:
Neither the database manager nor the *instructor knows* the answer.
Neither the instructor nor the *students know* the answer.

SUBJECT-PRONOUN AGREEMENT

RULE Pronouns must agree with their subjects. The words *each* and *someone* must use *his or her.*
For example:
Incorrect: *Each* subject in the study recorded *their* daily intake of calories.
Correct: *Each* subject in the study recorded *his or her* daily intake of calories.
To solve the cumbersome "his or her" problem, use the plural whenever possible.
Correct: *Subjects* recorded *their* daily intake of calories.

SKILL BUILDER HOW TO WRITE EFFECTIVE DOCUMENTS

Major writing tasks (reports, proposals, business plans) can be daunting. The following suggestions should take some of the sting out of getting them done.

Ask questions and establish objectives for every document. Anticipate the questions your readers will ask and make sure you cover them in your document. You may even want to use these questions as subheadings.

Ask yourself:

- Why do I need this funding/item/policy/change/new employee?
- What will our department/organization gain from this project? Think numbers.

Ask others:

- What do you think about this idea?
- Would a change in our policy help your department/job/organization?
- What do you want/need for me to cover in this proposal? (Ask both colleagues who may offer good ideas and superiors who may read the finished document.)

Get started. The hardest part of any writing project is typing the first word. No time? Make an appointment with yourself to do this task. Spend fifteen minutes per day brainstorming about the project. Make notes on a legal pad, on your computer, or in your cell phone. Writer's block? Occasionally even professional writers get stuck. To overcome writer's block, try the following:

- Jot down something, even if it seems unimportant. This free writing triggers ideas, stimulates thinking, and helps you manage the concepts you want to cover in your document.

- Use mind mapping, Venn diagrams, algorithms, outlines, or lists to get started. We actually started this book with a brief outline of potential chapters on a legal pad.

- Read. Continually refer to the text you have previously written, the research you have gathered in preparation to write, or a well-written document that motivates you. These activities will help keep you on task, stimulate thinking, and refine your ideas.

Think conceptually as you write each section. For example, how does your idea fit into the organization's short- and long-term plans? How will your idea coincide with market and industry changes? Refer to your initial questions to make sure you meet the objectives of the document.

Allow enough time for the project.

- Let the document rest. If you have time, set the project aside for a day or two so you can edit it with a fresh perspective. If you have a pressing deadline, take a short break to work on another task, or change locations to edit.

- Ask others to proofread. Ask for advice from both supporters and nonsupporters on your project. Make sure at least one proofreader is a good editor.

- Read your writing out loud. We often "hear" mistakes that our eyes didn't pick up.

- Print a hard copy. Even if you're on a tight deadline, you may find it helpful to print and proof a hard copy rather than editing from your computer screen.

WRITE FOR CLARITY

Great writing doesn't have to be complicated. In fact, the most effective writing is usually quite simple.

45. Begin documents with *thank you* and use *please* before making requests. (*Thank you* and *please* are still magic words.)

When writing letters and email messages, many people struggle with how to begin. Rather than writing a bunch of irrelevant fluff at the beginning of your letter, try thanking the other person for something.

For example:

> Dear Ms. Lobraico,
> Thank you for contacting me about the Klein Project . . .

Making a request can be awkward, especially when writing to someone you don't know. That's when *please* can open the door for you.

For example:

> Would you please proof this report and send back comments by Wednesday? (Note that this example is in the form of a question with specifics, making it gracious but firm.)

46. Write out acronyms, abbreviations, and contractions. Almost every workforce has its own jargon. To avoid confusion, write the way newspaper reporters do: spell out all names the first time with the acronym noted in parentheses and then use the shortened version after that. For formal business documents, avoid contractions. You may have noticed that we use contractions in this book. *That's* because we wanted it to have a conversational tone.

47. Replace phrases of weak adjectives, adverbs, and prepositions with strong action verbs.

For example:

> *Instead of*
> After getting the directions, we quickly got down to the business at hand and got the project under way.
> *Write*
> After getting the directions, we *plunged* into the project.

WRITE TO BUILD CREDIBILITY FOR YOUR IDEAS

Thanks to continual improvements in technology, most laypeople have twenty-four-hour access to scientific information online and most scientists have a paperless trail of records following them. For these reasons, scientists in all areas must be judicious when explaining and documenting their work. Take the climate researchers who, in late 2009, attempted to manipulate data and tried to cover up the emails that revealed the deception. They essentially killed their own credibility with poor communication and the appearance of impropriety. Some scientists may argue that the laypeople who touted this transgression in the media don't fully understand the science: all the more reason to offer clear, precise, and accurate science writing. Perhaps the biggest failure of poor writing is not seeing the need to merge science news into the mainstream.

Solid writing is essential in establishing credibility in all science and technical fields, from explaining a new technology to clarifying a protocol. When you're able to explain your idea clearly to a specific audience, you show them that you have a command of the material and that you understand how they will interpret it.

Yet often engineers, computer technology developers, health care professionals, and other science specialists never receive proper training in writing—or they use jargon only a small group of specialists understand. When these professionals are called upon to draft a proposal or a paper, they flounder in the unknown, agonizing over the deadline, finally producing a hastily drafted document that fails to reflect their knowledge, commitment, and passion for their work. If you want to share your research and ideas with the world via academic journals, the next skill builder is the perfect place to get started.

SKILL BUILDER HOW TO INCREASE YOUR CHANCES OF GETTING PUBLISHED IN A PEER-REVIEWED JOURNAL

The most important research in the world is valueless if it's not communicated to other people. Many of us dream of being published in *Nature* or *Science,* but you'll save yourself time and piles of rejection letters if you have realistic expectations about your manuscript. Below are some tips to help optimize your chances of getting your manuscript accepted.

Read one or more of the excellent books on writing research papers. There are many books from which to choose. (See the recommended resources at the end of this book for our favorites.)

Select the appropriate journal for your manuscript.

- Consult the journal's author guidelines and editorial policy regarding aim or scope.
- Look through recent issues to see what other papers are being published by the journal. Who are the readers? What are the topics? Is your topic one in which readers would be interested?
- Look at the editorial board. Do you know any people on it? If so, ask what they think. See if anyone you know has been a referee for the journal. Ask their opinion.

Read the journal's ethical guidelines carefully to avoid any misunderstandings around issues such as authorship, conflict of interest, plagiarism, ghostwriting, and bias in reporting. Recently these topics have come under fire, and you don't want to be caught in any inadvertent, unethical behaviors.

Ask colleagues with strong English-language skills and experience publishing scientific papers to read your manuscript and provide constructive criticism.

In an article titled "Eleven Reasons Why Manuscripts Are Rejected," failure to conform to the targeted journal (that is, not knowing your audience) is number two and poor writing is number three.[2] In a separate article, published in *Academic Medicine,* the editors note that the main strengths of accepted manuscripts included "(1) importance or timeliness of the problem studied, (2) excellence of writing, and (3) soundness of study design."[3]

Poor-quality writing loses credibility because it lacks clarity, precision, and logic. Inexpert writers often shroud their messages with hedging words and phrases. This style of writing weakens their argument, calling into question the validity of the science or logic, rather than giving credence to the research or idea communicated.

Researcher and linguist Marie-Claude Roland of the French National Institute for Agricultural Research notes "four distinctive features of hedging: impersonal style, passive voice, modal verbs, and the descriptive approach."[4] Hedging perpetuates unclear papers and unpersuasive scientists. Consider the following alternatives.

48. Choose active verbs. Active verbs are exactly what their name implies: active. They bring action to the sentence and make it stronger. To accomplish this writing style, try these two tips.

■ **Avoid forms of** *to be.*

These include *am, are, be, been, being, is, was,* and *were.*

Occasionally these words are the best choice. Use them sparingly as long as the doer is at the beginning of the sentence.

For example:

> *Instead of*
> A decision will be made by the *board.*
> *Write*
> The *board* will decide.

The active voice requires fewer words and it's the way we generally speak: subject, verb, object. The doer (the board) is at the beginning of the sentence. The word *decision* changes from a noun to a verb (*decide*), which gives the sentence action and therefore strength.

By the way Many science and technology professionals tell us that the reason they write in the passive voice (for example, *a study was conducted*) is to avoid overusing the personal pronouns *I* or *we*. Let us give you permission now: it's okay to use personal pronouns. What if, after twenty years of painstaking research, you and your colleagues discovered a cure for cancer? Would you say, *A cure for cancer was found*?

▪ **Replace nouns that end in** *-ance, -ing, -ment, -sion,* or *-tion* with corresponding verbs.

For example:

Instead of
We need a *stabilization* of the market before we can have a *modification* of our sales approach.
Write
The market needs to *stabilize* before we can *modify* our sales approach.

Using active voice empowers your writing because the reader doesn't have to reread or search for the information. It comes in a logical order, comparable to the way we speak.

When you write letters asking others to take action, avoid passive voice as it will make you sound patronizing. For example, many people write requests with *Your help is greatly appreciated in this matter.* Yet these same people would never say that out loud because it sounds stuffy. Changing the statement to active voice sounds much more gracious: *Thanks for your help.* Notice how we dropped *in this matter* because it didn't add value or clarification to the statement.

By the way Generally speaking, you don't need to thank people *in advance.* A simple *thank you* will suffice and sound less condescending.

49. Avoid modal verbs that cast doubt on your claim. Modal verbs are "helping" verbs that modify the main verb by changing its certainty or value. Some modal verbs that take away from the primary verb by casting doubt are *can, may, might, ought to, should,* and *would.* These helpers actually hinder writing because they are indirect.

When you write using direct words, you appear more involved, and thus more accountable, in your claims. Instead of writing, *This outcome implies that we **may** need additional research,* write, *This outcome implies that we **will** need additional research.* If you did the research, own it.

50. Select words carefully. Writing with density takes practice. Many of us were taught as early as primary school to add rather than delete words. Many instructors assigned a specific length to papers, teaching us to add unnecessary words. They also rewarded us for using descriptive words (adjectives and adverbs) rather than active verbs. Our papers were graded on length, not clarity. Today, less is more. Consider the raving success of tweets. Descriptive words and phrases have their place in writing, even in science, technology, and medicine. The key to using them is to make sure they are necessary and relevant.

▌ SKILL BUILDER HOW TO WRITE WITH DENSITY

Writing with density takes practice but the end result is worth it. The first step is to allow time to edit and rewrite your documents. During this process, ask yourself, "Does this word/phrase/sentence add value to my point?" If the answer is "no," take it out. Eventually you'll write with density without even trying. Here are three simple examples.

Shorten the verb phrase.

Instead of
The document was rewritten by Sam in order to improve its clarity and conciseness.
Write
Sam rewrote the document to improve clarity and conciseness.

In the shorter version, we changed the verb from a string of words to just one word, giving the sentence greater density. We also moved Sam (the doer) to the beginning of the sentence, creating active voice.

Remove words that don't add value.

Instead of
In fact, many times we spend extra time sending messages back and forth on email when we should have made the five-minute phone call to get the answer in the first place.
Write
In fact, often we waste time popping back and forth on email when a quick call could get the answer.

Reduce the number of words in a phrase.

Instead of	*Use*
a limited number	one
an overwhelming amount	most
a proportion of	some
a sizable percentage of	many
in the near future	soon
is at variance	differs from
is of the opinion	believes
make a statement saying	say
to a certain extent	in part
at this point in time	now
whether or not	whether
owing to the fact that	since
in spite of the fact that	although
unaware of the fact that	unaware
is in violation of	violates
to a large extent	largely[5]

51. Avoid jargon, clichés, and too many acronyms. Your writing will start to look and sound comical if it's full of overused phrases and unknown abbreviations. Some of our banned favorites: *ball park figure, circle back, deep dive, drill down, face time, grow the brand, hardstop, impactful, in the loop, leverage, on the same page, rightsizing, siloing,* and *synergize.* Some overused acronyms we see are *AOB, ASAP, BTB, COB, OOO,* and *ROI.* When using them, be sure you know your audience. New and often incorrect usage is always flourishing. If you have any of your own favorites, email us. We'd love to hear from you.

WRITE TO BUILD SUPPORT FOR YOUR IDEAS

Good writing can be very persuasive. First, others are more likely to read your document if it's easy to understand, makes a logical argument, and uses correct grammar, punctuation, and language. Second, you can use benefit statements, sentence structure, and word choice to persuade your readers to support your idea. The following tips present specific ways to persuade others and negotiate a deal. You can use them in both speaking and writing. Remember this caveat: the audience is the most important consideration.

52. Organize your document in a logical manner. Most people tend to write in chronological order instead of priority order. For example, first this event happened, then we responded with this thing that didn't work, so we changed course and tried something different, and finally we have this great result. Yet few of us are willing to wade through the details when we read if we can just flip ahead to the conclusion. Think about the order in which you read scholarly articles: title, abstract, conclusion, and, if the topic is worthy, introduction, methods, and results. You jump ahead to get to the punch line and go back to read the details if you need them. Your reader wants the same thing: easy access to the final point.

The Center for Plain Language website offers great advice for organizing your document to increase comprehension.[6] We have expanded their ideas with practical comments below.

- **Put the main message first.** Think topic sentence. You learned in grammar school to put the topic of the paragraph at the beginning and so did your reader. Put it where the reader expects it.

- **Divide your material into short sections.** Think journalistic style. Everyone enjoys a mini break when reading, which is one of the reasons news stories have lots of short paragraphs.
- **Group related ideas together.** Think Venn diagram. When you can grab multiple concepts from various sources and experiences, and weave them together, you show your reader that you have a command of your topic and can incorporate critical thinking skills.
- **Put material in an order that makes the best sense to the reader.** Think logical not chronological order. Offer the conclusion first, and then capture the audience with your methods, examples, and supporting evidence.
- **Use lots of headings.** Think visual appeal. Headings naturally lead us to read the details below them.

53. Use benefit statements to give your points credence. Benefit statements offer the reader the WIIFM (What's In It For Me?). An easy way to use benefit statements in your writing is to have a "you" focus. Another way to show the benefit is with an if/then statement.

For example:

If we reduce spending by 20 percent, *then* we can afford to hire another product manager.

54. Place the most important information at the end of the sentence in the stress position.[7] Researchers agree that the end of the sentence is the stress position because it's the last thing the reader (or listener) comprehends, and it sticks.

Example A:
The drug has significant side effects *but is highly effective.*

Example B:
The drug is highly effective *but has significant side effects.*

Presented with these two statements, more people say they would take the drug in Example A. Why? The emphasis is on effectiveness, not side effects.

55. Read your work aloud so you can hear how it sounds.

56. Use short sentences for important points. One sentence we use several times in this book is just one word: Practice. It has power because it's short. Amid paragraphs of long sentences, the short one always stands out. As William Zinsser points out in his excellent book, *On Writing Well,* "Among good writers, it is the short sentence that predominates. And don't tell me about Norman Mailer—he's a genius. If you want to write long sentences, be a genius."[8]

57. Use repetition and transitions to weave your paragraphs together. As readers, we look for visual cues that will aid in comprehension and retention.

For example:

In our section on dealing with difficult situations, we offer this tip:

Follow up to make sure everyone is satisfied with your actions. If you're not the person who can solve the problem, it might seem unnecessary to follow up. On the contrary, that's exactly why you should follow up. Both the complainer (who may be a customer, colleague, or boss) and your peers will see you as a problem solver. As we said earlier, this is the ultimate tag you want associated with your name.

Note the repetition
Notice how we strategically repeated the key phrase, *follow up.*
This repetition led you into our main point: why you should
follow up (also the benefit statement).

Note the transition
Notice how we wrote, *As we said earlier,* at the beginning of the
last sentence. This transition ties the pieces of the book together
and triggers you (the reader) to assimilate the new information
into what you already know.

58. Choose persuasive words to make your points. Over the
years, marketing experts have made lists with the most persuasive words in the English language. Nobody seems to know the
original sources of these words but common sense dictates that
they must be persuasive. We chose words from a list developed
by a group of education experts at the University of Maryland,[9]
which we have edited to suit you. Choose wisely from this list as
your goal is to persuade readers, not scare, offend, or threaten
them to support your idea. These words apply to both writing
and speaking.

In support of

accurate	advantage	always	best
certain	compelling	confident	convenient
definitely	effective	emphasize	guarantee
most	most important	popular	profitable
recommend	should	superior	trustworthy
worthwhile			

Against

aggravate	confusing	damaging	displeased
fraudulent	harmful	inconsiderate	inconsistent
inferior	irritate	offend	provoke
severe	shameful	shocking	terrible
unreliable	unstable		

USE VISUAL CUES

An easy way to make both online and print documents readable and appealing to your audience is to use visual cues. These cues are especially helpful in blog posts, websites, and print materials written for the lay audience. Darren Rowse of ProBlogger.net says one of the most important things you can do when writing a blog is make sure that it's visually scannable. Can a reader who isn't familiar with your blog capture the main point after fifteen seconds of reading?[10]

Visual cues do just what their name implies: the design of the document plays into the content. Here's how to adopt visual cues without formal design training.

59. Write subheadings for each section. Subheadings draw the reader into the topic by offering a prelude to what is coming. Subheadings are particularly effective in email messages because they're unexpected.

By the way Questions make effective subheadings. They draw the reader into the content, allowing you to write more persuasively. In selecting the type of questions, consider your audience.

60. Choose bold text for headings, important statistics, or other notable information. We have used bold text throughout this book. Did it catch your attention?

61. Add white space between paragraphs, sections, and other long strings of text. We can't emphasize this need enough. White space

- Provides a visual pause for your reader
- Gives the text around it greater importance
- Helps you highlight important points

The previous section is a case in point for white space. Notice how the bulleted list jumps out. This section and the rest of this book seem easy to read and manageable. White space helps accomplish this goal.

62. Change lists of items separated by commas into bulleted or numbered lists. In the white space example above, we changed a string of text into a bulleted list to make it easier to read.

Check out the same list of items without bullets.

White space provides a visual pause for your reader, gives the text around it greater importance, and helps you highlight important points.

SKILL BUILDER HOW TO WRITE BULLETED LISTS

Bulleted lists have even greater reading power when they are parallel in structure. Parallel structure means the first words in a list are the same part of speech (such as all nouns or all verbs) and, if verbs, the same tense (such as all present or past tense).

For example:

Here's a bulleted list without parallel structure.

Our company has become more eco-friendly by starting the following programs:

- An in-house recycling system
- Carpool for employees
- Turning off lights to conserve energy

In the above example, the first word after each bullet is a different part of speech: *an* is an article, *carpool* is a noun, and *turning* is a gerund (an -*ing* verb that acts as a noun). Because the list does not use parallel structure, the items seem disconnected.

Here's the same information in a bulleted list with parallel structure.

To become more eco-friendly, our company has

- Created an in-house recycling system
- Started an employee carpool program
- Conserved energy with our Lights Out! program

In the parallel structure example, the first word following each bullet is a past-tense verb. Try using verbs as the first word after each bullet in your list whenever possible. Remember, verbs drive the English language. Verbs are the only part of speech that can stand alone and still represent a complete thought, for example, "Run!"

Notice how the bulleted list with parallel structure rolls off your tongue when you read it aloud. Parallel structure makes sense to the reader because it flows rhythmically and lends a logical sequence to the list of items.

By the way Writing with bulleted lists helps you avoid overusing the personal pronouns *I* and *we*. In the following example, *we* appears once instead of three times.

We approached this problem to determine how to

- Allocate product and reject streams from the treatment unit
- Use bypassed freshwater
- Satisfy multiple process sinks with their respective flowrate and purity requirements

63. Avoid justified text; use flush left instead. Many word processing programs add space between the letters of words to make the text reach the other side. These stretched-out words are more difficult to read. Like most books, this one has justified text because that is the standard in book publishing.

64. Start new paragraphs often. Nobody enjoys reading a full-page paragraph. Starting a new paragraph gives your reader a visual break as well as a cue that the content is shifting.

Start a new paragraph when introducing

- A subject change
- A new point
- Important information

SKILL BUILDER HOW TO WRITE A RÉSUMÉ OR CV

To get a job, you need to get an interview. And to get an interview, you need to get in the door. To get in the door, you need a powerful résumé or curriculum vitae (CV). The question is, Which should you use: résumé or CV?

A résumé is a one- or two-page summary of your education, work experience, and skills. It is used most frequently in business situations. A CV, in contrast, tends to be longer and includes a summary of your educational and academic backgrounds, as well as teaching and research experience, publications, presentations, awards, honors, and affiliations. A CV is used primarily when applying for academic, education, scientific, or research positions.

You should think of your CV or résumé as a marketing tool. It will be the first opportunity for you to convince the person reading it that you can solve the needs of the company or organization. If it's hard to read or full of errors, it's likely to end up in the circular file. Here are some important tips to help get your CV or résumé read and acted upon.

Center your name and complete contact information at the top of your document. Include full name, home address, phone numbers, fax numbers. A word of caution: never use a current work phone or fax machine for a job search.

Avoid irrelevant information such as age, sex, hobbies, marital status, religious or political affiliation, and sexual preference. Rather than help you, this information may in fact hurt you. Leave it out.

Put your most recent job or position first. Think of your CV or résumé as an interesting story that has naturally led to the position you're seeking.

Use clear, concise, correct English. Use action verbs and parallel construction, e.g., *achieved, developed, directed, engineered, implemented, mentored.*

Focus on achievements instead of just activities and responsibilities. Too many CVs and résumés look more like job descriptions. Don't just write, *Responsibilities included . . .* and then make a long list. Pack the document with precise examples and facts about you, your impact on your recent company or group, and why these skills or experiences will translate well in the new position. Have a story for every skill.

Custom-design the CV or résumé for each job or position you want. One size won't fit all. Be sure to use the keywords found in the job description and then make a point to spotlight specific experiences or skills that the potential employer is seeking. Highlight those skills or experiences that make you a perfect fit for the position being posted.

Be accurate and absolutely truthful. Never embellish or lie. Most employers today check everything on a CV or résumé, including schools listed and places of employment, as well as dates.

Design with simplicity and elegance in mind. Make your document appealing and easy to read. Use bullet points, headings, action verbs, and plenty of white space. Choose a typeface such as 12 pt Garamond or Times Roman. You want your CV or résumé to stand out from the dozens in the pile.

Avoid clichés and overused words and phrases such as *dynamic, innovative,* and *motivated.* Such words are among the most overused on résumés, according to networking site LinkedIn, which analyzed the profiles of millions of U.S. users to come up with a top-ten list.[11] Also seen too frequently: *entrepreneurial, extensive experience, fast-paced, problem solver, proven track record, results-oriented,* and *team player.*

Proofread, proofread, proofread. Then ask someone to proofread for you. Grammatical or punctuation errors can put the kibosh on any hopes for an interview.

After you draft that fabulous CV or résumé, you'll need to send it out with a cover letter. Often our audience members ask, When should I include a cover letter? The answer: think of your CV or résumé as a package that you are giving away. If you were to give a package to someone in person, you would not necessarily include an enclosure card. If you boxed up and mailed the package, you would likely include a brief note or card. Thus, if you are going to hand your CV or résumé to someone, and you have an opportunity to explain it in person, no cover letter is required. Otherwise, assume the materials will be passed along without explanation and therefore need a cover letter.

WRITE A COVER LETTER TO GET AN INTERVIEW

Your cover letter should tell your reader something that your CV or résumé doesn't say. Use it to highlight your talents, such as things you learned in your last position and the skills you have cultivated. Make sure your cover letter

- Is specific and custom-designed to the job and the reader
- Explains why you are qualified for the position
- Tells what talents you bring to the organization
- Asks for an interview

> *By the way* Avoid addressing your cover letter, "Dear Sir or Madam." If the job description does not list a contact person, call the organization and find out the name of the person to whom the letter should be addressed. Be sure to (1) ask for correct spelling and pronunciation of the person's name, (2) verify the person's title, and (3) thank the person who helped you get this information.

If you get the invitation for an interview, here's how to go through the process and position yourself to get the job.

SKILL BUILDER HOW TO SUCCEED IN AN INTERVIEW

Interviewing for a new position can be a nerve-wracking and grueling process, particularly for people in the fields of science, technology, and medicine where "selling oneself" has generally never been part of the "program." The good news is that strong interviewing techniques are a communication skill that anyone can learn with continued practice. Below are tips to help you remain calm and collected while presenting yourself more effectively and winning over interviewers.

Do your homework and be fully prepared. Know everything you can about the company and the people with whom you'll be meeting. Study the company webpage, scour recent business articles about the organization, and be prepared to intelligently discuss recent company events. Use social networking sites like LinkedIn and Facebook to find out details about professional organizations to which the interviewers belong. Finally, do a literature search using the last names of the people with whom you will be meeting and be prepared to discuss their research or areas of interests.

Dress for success. First impressions count. You want to project an image of a well-organized, well-prepared, motivated worker. Make sure your clothes are clean, comfortable, and well fitted. Even if the workplace is business casual, it is better to be overdressed than underdressed in your interview. Men should wear a dark suit, solid shirt, and simple tie. Women should wear a dark suit, simple jewelry and accessories, and conservative makeup. Pay attention to details: avoid scuffed shoes, unkempt hair, and chipped nail polish.

Practice good communication skills. Be sure to read chapter 6, "Serve," for more tips on communication skills and professionalism.

■ **Be aware of your body language.** More than 50 percent of what we communicate is through body language—not speech.[12] Eye contact with everyone interviewing you is essential.

- **Have a firm handshake and remember your manners,** especially if you are having a meal with your interviewers.
- **Practice active listening.** This is your chance to connect with the interviewer by practicing the active listening skills you learned from chapter 2, "Listen."
- **Adapt your style to the interviewers'.** Pay attention to details of dress, office decorations, general décor. These will give you clues on how to tailor your communication style.
- **Encourage the interviewer to share information about his or her position.** Have specific questions for each person with whom you will be meeting to show you've done your homework and demonstrate your interest in him or her as well as the organization.
- **Take notes** so that you will be able to ask questions and refer to them later when you're writing thank-you notes.
- **Try to strike a balance between confidence and humility.** Nobody likes exaggerated boastfulness or excessive obsequiousness.
- **Finally, be yourself.**

Prepare a portfolio of work-related samples. Include your CV or résumé, copies of articles you've published, previous employee evaluations, and anything relevant to the job. Have enough copies to give to everyone you will be meeting.

Be precise and specific in your answers. Even if the interviewer is vague and says, "Tell me about yourself," cite specific examples and scenarios that highlight your abilities, talents, and skills—ones that will make you a perfect fit for the position. Keep in mind that not all interviewers are good at interviewing. Many don't even prepare for the interview. As the candidate, it's your responsibility to see that even the worst interviewer not only gets the information he or she needs from you, but also leaves the interview feeling that you're the one for the job.

Be ready for the tough questions.

▪ **Why did you leave (your last job)?**

▪ **What is your greatest strength/weakness?** The "weakness question" is one of the trickiest to answer. Answering too honestly about your shortcomings can be as damaging as answering in a way that implies you have no faults at all. Try not to focus too much on the word *weakness* but rather on areas for *professional development*. Use the opportunity to talk about ways in which you hope to mature professionally in your career.

▪ **What would your former supervisor say about you?** Be careful here. You never want to say anything negative about a current or former boss. Highlight some of your strengths instead.

▪ **What is your greatest achievement?** Look through your résumé and have a brief story prepared for each skill or talent listed. Make sure your answer relates to the job to which you are applying.

▪ **What did you do from _____ to _____** (hole in your CV)? Be honest. If you took a year off to be a surfer dude, say so. But make sure you explain how that year developed you as a more mature professional person capable of doing the job at hand—not just as a surfer.

Remember, the key to a great interview is to stop talking after you answer the question completely.

Have well-prepared answers for the "Tell me a time when" question. Many interviewers today like to ask this question as a way of determining your flexibility and how well you handle stress, change, and disappointment. Again, look through your CV and past performance reviews. Then generate a list of examples where you can tell a compelling, but clear and concise, story that demonstrates how you solved a problem, took the lead on a project, showed a level of leadership, or bounced back from adversity. Since the first grunt around the fire, people have loved stories. Make stories work for you. But don't forget good storytelling takes practice, practice, practice.

Be prepared to give a brief presentation in front of a small group. More and more companies are recognizing one of the quickest ways to eliminate candidates is to see them present. If you're looking for a job in the pharmaceutical or biotechnology field, it's very likely that you'll be asked to give a presentation in front of several people outside your discipline. We urge you to carefully read chapter 4, "Present." Many excellent, well-qualified scientists have not been hired into industry because their presentation skills were poor.

Summarize why they should hire you at the end of each interview. Reiterate your strengths. "My understanding is that you're looking for somebody who can do A, B, and C, and I think you'll agree I can bring all of these skills to the organization." It's important to leave the interviewer with two lasting impressions: (1) you are sincerely interested, and (2) you have everything that they are looking for in a candidate.

Follow up by sending a thank-you note. This is a must. A thank-you note not only shows respect; it also shows that you are truly interested. It also gives you a second chance to

- Include information you forgot to mention
- Reiterate important information
- Correct any impressions you think may have been misinterpreted
- Ask for the job

The thank-you note can be an email, but you should be sure to send a separate one to each person with whom you met. Personalize each email by highlighting some part of the conversation you had with that particular person. Proofread, proofread, proofread.

WRITE EFFECTIVE EMAIL AND TEXT MESSAGES

Thanks to technology, two of the most common forms of writing today are emails and text messages. Don't let the fact that you may write a large number of these daily keep you from making them well written.

Before you begin the exchange of email or text messages, ask yourself: "Would it be more effective for me to call or stop by and visit with this person instead?" Many businesspeople complain that they are inundated with emails. When we ask our audiences how many emails they receive per day, some of them say over 150. At one minute per email, that's two and a half hours of emails. It's no surprise that, when reading email, their goal is to act and delete. For this reason, your emails to others need to be worth reading.

65. Give others a protocol for sending emails. Let them know that you're inundated with emails. Tell them what information you wish to receive by email. Ask your subordinates for a weekly synopsis rather than a play-by-play of their work.

66. Use emails for brief communications: requesting information, setting up meetings, answering quick questions, thanking someone, delegating a task.

67. Be specific in the subject line. This will help you store and locate emails quickly. Plus, the recipient will be more likely to read it.

68. Put all of your contact information at the bottom of all emails so people can readily reach you by email, phone, or fax.

69. Specify where to send requested information. Does the reply need to go to you, your assistant, or a colleague? Most people will hit "reply" and automatically send their response to the email sender. Help direct the reply by copying the message to relevant colleagues and inserting an email address in the body of the message. For example, "Will you please send the report to Alex Farnham, alex.farnham@abc.com, by Friday, August 21?"

70. Cover only one point per email. Many people retain and respond to the first or last point only, ignoring your other points. If you must cover several items, write them in a list format, with numbers, bullets, or dashes in front of each point. This encourages the recipient to respond to each line item.

By the way If someone asks multiple questions in an email, cut and paste the email in your reply and answer each question separately in a different color font.

71. Read the most recent emails first, and then move down the list before you respond. Make sure you have the latest information before you respond to emails. If you read and respond to emails in chronological order, you may miss out on your colleagues' responses. The beauty of this technique is that you may not need to respond at all.

72. Clean up strings of emails by deleting the extras. If possible, when you reply to a message, delete the preceding message that your email program automatically included. Passing the same message back and forth without updating it dilutes the final product. Of course, keep the details (or a separate copy of prior messages) if you need them for documentation.

By the way As your relationship and/or project progresses, your email messages with your colleague or client may take on a less formal approach. For example, you may decide to omit the salutation, "Dear _____," and go straight to the point. This is okay as long as the messages are mutually understood. Always try to adapt to the other person's style.

73. Don't "reply all" unless the message is relevant to all. This applies to things like invitations to workshops or meetings. The other invitees don't need to know if you'll be attending.

74. Be aware of your organization's compliance rules. You may need to save email messages to a specific folder or server rather than deleting them.

75. Avoid putting anything unprofessional (jokes, cartoons, personal information) in your emails. If nothing else, you're wasting your time and your colleagues'. Further, there is potential for career damage should your boss end up receiving them.

76. Avoid using emoticons (smiley faces, frowns, and other graphics) in your emails. Emoticons are fine for personal messages but unsuitable for business. Ask yourself: does this add value and professionalism to my message?

77. Use your work email (and other electronic devices) for work only. This tip sounds so obvious we almost forgot to include it. Accessing your personal email at work isn't just unethical; it could hurt your career if your friends send you inappropriate messages.

78. Assume all email messages will be forwarded, saved, and seen by others. These (as well as text messages, voice messages, and computer files) are legal documents. If your name is on the message, there is a direct connection to you.

By the way Don't be afraid to pick up the phone and call the other party instead of responding by email. Call when

- The response requires some discussion or brainstorming
- The other party is confused about the previous message or topic
- The sender has initiated a financial negotiation
- The topic is sensitive and the message does not need to be forwarded
- The information should not be put in an email

79. Send copies to others only when they need or request them. You don't need to document your every move by sending a carbon copy of every email to your boss. Instead, ask your boss if an end-of-the-week synopsis will do. Your boss will be more likely to read and appreciate one message.

80. Use complete words for business text messages. While acronyms in text messages are clever and often universally understood, using a single letter or symbol to represent a word leaves room for misinterpretation and legal issues. For example, wouldn't you feel awkward if you sent a text to your boss with LOL meaning "laugh out loud" but he or she interpreted it to mean "lots of love"? When you first communicate with someone via text message, be sure to sign your name at the end in case you're not identified by caller ID.

By the way If possible, include the main information in the body of an email message rather than as an attachment. Your recipient will be more likely to read it, especially on a portable device.

WRITE CONVINCING REPORTS AND PROPOSALS

To be effective and persuasive, reports and proposals require research and time (and a separate book from us). Make sure your readers will want to read these documents by applying the following writing techniques.

81. Match your proposal to your purpose and your audience. Know your audience. Know your purpose. Use these points as the foundation of your document:

- Strive to communicate, not to impress.
- Be clear. Avoid jargon.
- Make your proposal visually pleasing.

82. Make sure each point or goal can stand alone. Some people who read your proposal will skim-read. They will hit the headlines and delve into the sections that are most relevant to them. If each section can stand alone, then your reader will be able to capture your point in just one section. Make your sections stand alone by repeating key phrases, not by being redundant.

83. Offer specific dates, names, and data to support your points. Being specific removes the margin for error. The best data to support your points should come from your organization. Be sure to reference your sources.

For example:

> According to Kris Wilkins, our VP of Finance, the budget for next year must decrease by 3.5 percent or $526,000.

84. Plan time to edit. One of the biggest mistakes people make in writing is waiting until the last minute rather than allowing time to edit their work. The next time you read a great novel, notice the many editors thanked in the acknowledgments. Great writers know that great writing comes from great editing.

85. Make it error free. You don't have to know every grammar, syntax, and punctuation rule to be an effective writer. As one of your grammar school teachers probably told you, "Look it up." Don't rely on Spell Check and Grammar Check alone. There are great websites and style manuals for writers. (We list several such resources at the end of this book.) Double-check your facts also. The last thing you want is for your reader to catch an error.

By the way Don't get tangled up in nonproductive back-and-forth emails and voicemails. Sometimes you just need to talk in person. If it is geographically feasible, a face-to-face meeting can be the lifeline to a project. Times to initiate an in-person conversation or meeting are when

- Your emails or phone calls back and forth are not moving the project forward
- Someone does not understand the scope of the project or is not meeting deadlines
- You and the other party are continually having differences of opinion in meetings or conference calls
- You are the chair of a committee or leader of a project to which another person is not contributing adequately or is imposing a negative attitude on the rest of the group

SKILL BUILDER HOW TO EDIT YOUR OWN WORK

Being able to edit your own work is a necessity. The key to effective editing is to start writing early so you have time for your document (and you) to rest before it's due. The following steps stem from the tips offered in this chapter.

Ask yourself the following questions.

Is the document

- Reader-based?
- Purposeful?
- Clear?
- Concise?
- Well organized?

Check spelling and grammar. Use Spell Check (or read the document backwards) and Grammar Check. Don't rely on these alone, however. Proof your spelling and check your grammar.

▌ **Check organization.** Read the entire document without stopping to make edits. Does it flow?

▌ **Print a hard copy and read it out loud.** The errors jump off the page when you hold it in your hand and hear what you've written.

▌ **Underline any sentences that you have to reread** and rewrite them more simply. Choose one idea per sentence.

▌ **Rewrite passive voice sentences.** Circle any *to be* verbs—*am, are, be, been, being, is, was, were*—and replace them with action verbs.

▌ Make sure the doer is at the beginning of the sentence next to the action verb.

▌ **Circle all nouns in a sentence that end in -*ance*, -*ing*, -*ment*, -*sion*, or -*tion*,** and replace them with corresponding verbs.

▌ **Consider changing lists of items separated by commas into bulleted points.** This gives your reader a visual break and draws attention to the important points.

▌ **Reprint the document and set it aside** to proof again later (the next day, if possible).

▌ **Ask a colleague to proof.** Choose someone who is a good proofer, and unafraid to make edits, and be open to receiving suggestions.

4 Present

Presenting is more than standing up in front of a crowd to give a talk. We present ourselves every day to everyone we encounter. From the moment you approach your workplace to the time you leave, you'll likely communicate with many different people, in many different ways. The most influential way you communicate is face-to-face, meaning how you present yourself and your ideas.[1]

It's no accident that this chapter is one of the longest in this book. We cover a lot of important skills here: communicating effectively in one-on-one interactions, organizing and delivering presentations, learning to design slides, and answering difficult questions. Mastering these skills translates into coveted leadership skills: being able to persuade others to adopt your ideas, negotiate for mutually beneficial outcomes, manage important projects and collaborate with the people working on them, explain complex data to lay audiences, and showcase your ability to think conceptually and predict trends.

Once again, planning and preparation are essential:

- Know your audience.
- Know your purpose.
- Know your subject matter.

Speaking is a fundamental part of our lives. It's the first communication skill we learn. Because it comes "naturally," it seems that we don't really need to work at this skill. On the contrary, that is exactly why we need to cultivate it.

THE POWER OF POSITIVE COMMUNICATION

What you say, how you say it, and why you say it speak volumes. Make sure that the words coming from your mouth and the actions accompanying them truly reflect what you want to communicate. Whether you're communicating to your patients, your clients, your colleagues, or your boss, your success depends on structuring a clear message and delivering that message with confidence and conviction.

Many experts have written entire books on the power of positive thinking. We have all heard interesting success stories of the patient who recovered, against all odds, from a terrible accident; the IT student who dropped out of college to start a multimillion-dollar business; the scientists who performed thousands of tests before discovering a breakthrough treatment. Positive thinking and the communication that accompanies it yield positive results. You can harness the power of positive communication, too. Make it your goal to communicate in a way that makes others perceive you as upbeat and positive. The following tips on how to speak positively may seem elementary yet they are among the most popular we offer in our seminars.

86. Speak to everyone you see. When you arrive at work or another destination where you'll see familiar faces, make eye contact, smile, and acknowledge others. This does not mean you have to stop for a ten-minute conversation with everyone you see. Rather, it means that you should offer a polite greeting as you pass. You'd be surprised how often the person you acknowledged in the hall turns out to be the schedule-keeper, decision-maker, or computer programmer you need to get a meeting set up, a job approved, or a lost computer file recovered. This small gesture will not only demonstrate your empathy and respect for others but encourage empathy and respect for you.

A surgeon once told us a memorable story about speaking to people. He said that he made a point of smiling and greeting everyone he encountered in his small town. One day, while shopping in a local grocery store, he spoke to a woman he didn't immediately recognize. He smiled at her and asked, "How are you today?" She answered, "I'm so much better since you operated on my foot!" Boy, was he glad he "spoke" to her. Numerous studies support the theory that dissatisfied customers (or patients) tell their disappointment stories to others. This woman would likely have been disappointed to find her surgeon ignoring her in public and would then have shared this story with others in her community. In a Wharton study of dissatisfied customers, only 6 percent reported their problem to the company, whereas a whopping 31 percent told friends, family, or colleagues about their experience.[2] Those friends, family, and colleagues add their twist to the story and it grows with each retelling, much like the "telephone game" in which schoolchildren in a row successively whisper a phrase to each other, and the last in line announces the final, altered version to the group.

By the way Speaking to people you pass in the hall at work inherently enhances security at your organization. When you smile and acknowledge people who are not supposed to be there, it says, "I can identify you." Surprisingly, humans can identify a smile from about three hundred feet away, the length of a football field.[3]

87. Select positive words. Rather than say, "I don't think that will work," try something positive such as "Let's try talking to them first." Also, when you disagree with another person, avoid using the word "but" for transition. For example, "I think Kathleen's idea could work, *and* I also think we should try Todd's."

88. Speak in specific terms. Being specific saves time. When you have a deadline, tell the other people involved in your project. Ask for the data "by Friday at 2:00" rather than "by the end of the week." Similarly, ask others to give you specifics: "I know you need this soon. Exactly when would you like me to send it?"

89. Avoid making sweeping statements. You can prevent yourself from making exaggerated statements by avoiding all-inclusive words such as "all," "every," "everyone," and so on.

90. Avoid sarcasm. It's okay to be self-assured. Just make sure you're open to other ideas and that your communication style reflects a positive attitude rather than cockiness. Remember, how you say things can sound sarcastic, even when that isn't your intent. For example, saying, "Dr. Lawrence's talk was certainly interesting" sounds sarcastic when the word "certainly" is emphasized.

Here's a true story about how a person's attempt at being clever was misinterpreted as sarcasm and backfired. A group of physicians asked their patients to complete a customer satisfaction survey for their practice. In the survey, several patients commented that one of the physicians was sarcastic and lacked bedside manner. The offending doctor was shocked to hear that he had received negative feedback since he had always thought his remarks to patients were clever and witty. To many people, he was bright and funny and some patients even enjoyed his teasing comments; however, other patients took his jokes to be insensitive. To his credit, he identified this problem early in his career, changed his communication style to accommodate a particular audience and setting, and rebounded nicely. How exactly did he do it? He began by paying greater attention to each patient's situation and communication style as well as cultural and educational background. He also tried to be more aware of signs of anxiety, fear, distrust, sadness, or frustration. In short, he recognized that not all situations called for humor and began to adjust his style to meet individual needs, thereby showing empathy and respect.

91. Adapt your communication style to your listener's style and the situation.

Two things you can easily adapt:

Mode. Ask your colleagues which mode of communication works best for them—voicemail, email, text message, or face-to-face—and choose accordingly. The most effective and persuasive is face-to-face. People respond better to communication by voice than by text.[4]

Technique. When communicating face-to-face, watch how the other person communicates nonverbally and mirror that behavior.

Mirroring requires really focusing on the other person. When we suggest mirroring to our audiences, they usually look appalled until they try it. Once they get the hang of it, the audience members can see many practical applications, including greater understanding of the other person's message. A few ways you can mirror another person's nonverbal gestures:

- Place yourself in the same position. Sit down if the other person is seated; stand up when he or she stands; lean in if the other person does so.
- Match the other person's pace of speaking in conversation.

If you're thinking, "Won't this mirroring just annoy the other person?" keep this in mind: psychiatrist Albert E. Scheflen noted in his research that people subconsciously tend to shift postures to mirror those of the people with whom they agree.[5] Obviously, if you overdo the mirroring, the other person will be offended, so be subtle.

SKILL BUILDER HOW TO PERSUADE OTHERS TO CHANGE THEIR BEHAVIOR OR ADOPT YOUR IDEA

Persuasion skills will serve you well in many everyday negotiations: asking a family member for a favor, influencing a colleague to support your idea, writing a winning proposal, helping a customer make an important decision, convincing a boss for a promotion. The key to being persuasive is to understand what the person you're trying to persuade thinks about the issue. Know your audience. As you read this book, you'll pick up other tips that relate to persuasion in both speaking and writing in various scenarios. In this skill builder, we offer five steps that are a great start for persuading others when speaking.

Plan and practice what you want to convey. Before you communicate with the person you're trying to persuade, make sure you research, plan, and practice what you want to say. During this process ask yourself, "If I were listening to this pitch, would I like this idea?" Keep in mind the importance of giving something up in order to gain something else and seeing your idea from the other person's perspective. Your goal is to offer enough information so the other person can make an informed decision.

PRACTICE

If you sound articulate, you'll appear thoughtful and persuasive. As with all good planning, thinking about your idea and the possible arguments against it ahead of time will help you cull the weak parts and fine-tune the strong ones. Practice is something we suggest repeatedly in this book. No amount of knowledge, research, or education can replace practicing or rehearsing what you want to say. Practice will also help you work through two important parts of your argument: delivery and word choice.

DELIVERY

We've all heard the joke that could have been funny if only the speaker had nailed the punch line. The way you deliver your verbal message affects the way others perceive it and you. Studies suggest that people who use hesitant language, such as "I mean," "um," and "you know," when they present a persuasive argument typically lose the negotiation.[6] Like a good joke, a good argument gets better with practice.

WORD CHOICE

When we develop an argument, or even just tell a story, we choose our words based on our education, experience, and background. Make sure the words you use reflect what you want to say and appeal to your audience. Many people have "lost" negotiations because they chose the wrong words. Choose words carefully when you're trying to influence others:

Instead of
"You should/need/ought to . . ."
Try
"Have you considered?"

Ask questions to build rapport and identify common ground. The more you know about a person and his or her perceptions of a project, the more you can understand how best to communicate and collaborate with him or her—saving you both time and misunderstandings. The key to building rapport isn't asking questions; it's listening to the answers. This is also a great way to find out what the other person needs. Ask about needs even if you think you already know what the other person is going to say. Remember, the other person's goal is to make you understand what he or she wants. Show that you do by listening actively.

Another important part of building rapport is acknowledging the position, education, experience, or investment of the other person. For example, "I know this project has been your baby for over five years." Show sincerity by making eye contact, smiling, and using appropriate tone. Remember, most people just want to be acknowledged for their contributions.

Establish a need by doing your homework. This seemingly minor step is actually the most important in persuading others. Most people don't usually buy things they don't think they need. The same applies to your idea. Find out what is important to the people you're trying to persuade. (You can do this by researching, talking to their colleagues, speaking to their office assistants, or simply asking them.) Then set up your case by integrating the "need." For example, "I understand from our conversation that you need . . . so here's how we can help you accomplish this."

After this step, you may need to let the negotiation rest for a few days, especially if the interaction was heated. Take cues from the other person and use good judgment, not your emotions, to determine your next communication or action.

By the way A great way to make a suggestion to someone is to word it in the form of a question. For example, "Would you mind if I contact the division manager and ask to speak to the group at their next meeting?" Asking permission shows respect and ensures that you're following protocol or "tribal regulations."

Focus the conversation on the benefits. Simply stated, benefits describe what the other person will gain. Many people try to persuade others by focusing on features rather than benefits. Here's the difference.

Feature statement:
"I think opening the company daycare an hour earlier will help cut down on employee tardiness."

Benefit statement:
"I think opening the company daycare an hour earlier will help cut down on employee tardiness *by allowing parents more time to get their kids settled before they have to go to work.*"

The benefit statement (in italics) tells the listener what the employees will gain. This is also a great place to insert a brief, relevant, and insightful story to support your point. An easy way to craft a benefit statement is to add, "What this means to you."

Close the conversation with a summary and, if applicable, a call to action. Summarizing the conversation is the best way to check your facts, show you were listening, and make sure both parties are on the same page. One way to create a call to action is to use an if/then statement. For example, "*If* I contact the members of my team, *then* will you get in touch with the other department heads?" An important nuance to consider in a negotiation is who holds the position of authority. If you're trying to persuade a superior, make certain you show respect for the person's position and ideas. Likewise, showing respect for subordinates will help further your goals in the future.

NONVERBAL COMMUNICATION

Nonverbal communication plays a significant role in how others perceive you. During the past forty years, many articles, books, and speakers have quoted the studies of Dr. Albert Mehrabian, Professor Emeritus of Psychology at UCLA. In his studies of interpersonal communication, Dr. Mehrabian concluded that there are three elements of face-to-face communication: words, tone of voice, and body language. His experiments specifically tested the theory that when people communicate face-to-face, they determine if they like or believe what the other person is saying based on all three elements. The most important finding is that when these elements are incongruent, the listener trusts the nonverbal gestures and tone of voice over the verbal communication. For example, if a person says, "I believe you," but avoids eye contact and has a closed body position (arms crossed, for example), the listener does not trust what the speaker said.[7]

Nonverbal gesturing also has cross-cultural implications. Charles Darwin theorized that evolution has programmed humans to express themselves universally.[8] This explains why a look of distress or happiness is probably similar in both Los Angeles and Nairobi. This research also supports our claim: nonverbal communication is a crucial component to sending effective messages. Check out these tips on using body language to enhance your message.

92. Make sure your nonverbal gestures match what you say.
When you communicate face-to-face, use eye contact, an open stance (no crossed arms), and appropriate hand gestures to ac-

company your points. The words you choose will carry more weight if your nonverbal gestures support them. For example, if you're explaining a new theory for your research to a group of financial supporters, you must appear confident. If you read from your notes, avoid eye contact, and fail to gesture, your listeners will think you're uncertain about the significance of your research, and they will not trust what you say or fund your project.

Establishing and retaining trust are crucial aspects of collaborating and negotiating. To help you incorporate these components into your interpersonal communications, we offer the following skill builder.

SKILL BUILDER HOW TO USE NONVERBAL GESTURES TO ADD VALUE TO YOUR MESSAGE

Many people have analyzed nonverbal gestures and their positive and negative implications. We developed this list to highlight gestures that will enhance what you say. Choose what works for you.

Eye contact. Making eye contact when you communicate with others helps you show empathy, gain trust, and retain information.[9]

Hand movement. When you gesture with your hands, you naturally look more interesting, have greater inflection in your voice, and enhance what you have to say with a built-in visual aid.

Body position. You can use your body to make big gestures such as moving toward your listeners to emphasize a point or to make small gestures such as leaning in to show empathy.

Stance. If you're standing when you speak, try to maintain an open posture (no crossed arms or ankles) with your arms bent at the elbows and your hands in front, ready to gesture.

Facial expressions. Most people show exactly what they are thinking with their facial expressions. Since the face has forty-four muscles and five thousand expressions, humans have developed distinct facial expressions to send specific signals to others.[10] Some positive facial expressions to use when speaking include lifted eyebrows, eyes wide open, and smiling. The best way to present positive facial expressions is to face the audience (even if it's just one person) with your head up so the listener can see and gauge your expression.

TELEPHONE COMMUNICATION

Telephone skills seem innate—how hard can it be to talk on the phone or leave a voice message? Try listening to voicemail from your colleagues a second time, to analyze for effectiveness. How could these messages be tighter? Did the caller give details or just waste time asking you to call back? How many times did you have to replay a given message before you understood what the caller wanted? Surprisingly, many people don't take advantage of the opportunity to use the telephone as a communication tool. These people are ill prepared when they leave messages, requesting a return call without specifying why or fumbling through a lengthy message while shrouding the request in unnecessary details. Even in "live" calls, these people take up extra time offering irrelevant details. Save yourself time and frustration by becoming an efficient telephone user. Many times the person on the other end of the line will match your style and become a better telephone communicator too.

93. Plan your message before you call and be prepared to leave a voicemail. The key to effective and efficient communication is planning. The one or two minutes you spend planning your call (or conversation) could save you several minutes of rambling to another person or, worse yet, getting cut off someone's voicemail for talking too long.

94. Speak slowly and clearly. Be sure to state your name, affiliation, and phone number. Because we know our vital information, we often say our name and phone number too fast for others to understand or jot down. Slow down.

95. Divide calls into introduction, body, and, conclusion. This tip sounds basic yet few people follow it. When you organize your call, make sure to include all important elements.

For example:

> *Introduction*
> "Hello, this is Mukesh Patel with SRI Technology. I'm calling about . . ."
> *Body*
> "I have two things I'd like to discuss." This statement offers an agenda of exactly where you're heading. It helps hold the listener's attention and keep the conversation on task.
> *Conclusion*
> "Thank you so much for your time. Would it be okay for me

to . . . ?" Use the conclusion to wrap up the conversation and make sure both parties agree on what actions to take next.

96. Get to the point immediately. It's okay to exchange pleasantries, for example, "Did you have a nice holiday?" Just make sure you move back to the purpose of the call quickly.

97. Practice before you call. Most people don't put enough stock in practice. Practicing helps you get focused to save time and, more importantly, sound articulate. At the very least, rehearse what you plan to say in your head, or jot down a few notes.

98. Offer specifics with the information you want, the best way to reach you, and when you'll call back. When you specify what you want in a message, you show the other person that you don't intend to waste his or her time. Plus, you expedite the process because you help prepare the other person for the return phone call.

For example:

Many people waste valuable time with nebulous phone messages that only request a return call: "This is Demetrice Owens. Please call me at 252-555-1234."

A better, more specific message (that doesn't have to be played repeatedly to get the details):

"This is Demetrice Owens from ABC Pharmacy. We're working on your refill but we have a question. Will you please call us back so we can fill it? Our number at ABC Pharmacy is 252-555-1234. Again, it's Demetrice Owens at 252-555-1234."

99. Train your telephone voice. Since you don't have to worry about eye contact, the telephone is a great place to fine-tune your communication skills.

- Replace vocal garbage such as "you know," "um," and other fillers with a pause. Nothing is more annoying in communicating than the "um." Most people don't even realize they say it.
- Smile when speaking to others over the phone. When you smile, you sound happier.
- Look at yourself in the mirror when you're on the phone. You'll sound better because you'll *see* how you sound.

100. Listen to your message before you press "send." Once you realize that you're being routed to voicemail, listen for the directions on how to send or delete your message. That way you can stop recording or rerecord if necessary. Every few months, send yourself a message and listen to how you sound.

By the way Many voicemail systems will interrupt when you press the # sign. This trick may help if you botch a message and need to rerecord.

SKILL BUILDER HOW TO TAKE "UM" OUT OF YOUR VOCABULARY

Over the years, we have given our audiences the "Um" Challenge: during a two-week period, practice not saying "um" (or "ah," "you know," or "like") in casual conversation. Once you master this, take the "Um" Challenge to the office and try it while speaking to your colleagues. If you take "um" out of your vocabulary now, it won't get in the way of your next important presentation.

Here's how:

Make a point not to say it. In setting this goal, you'll become aware of the many times "um" slips from your lips.

Make notes and rehearse before you speak. Most people say "um" because they lose their train of thought. If you have notes handy, you can refer to those instead of filling the void with "um."

Practice with a pause. Silence is a great attention-getter when presenting. It's also a great way to pause and collect your thoughts. You can avoid saying "um" by saying it in your head instead of out loud.

101. Ask permission before you put others on speaker phone. This is just common courtesy. Speaker phone is a great tool for a conference call. Otherwise, keep the phone at your ear. If you must use a speaker phone, make sure you introduce the caller to the other people in the room with you.

102. Assume all voicemails will be forwarded and possibly transcribed. Voicemails can be legal documents. Don't say anything in a voicemail you wouldn't say in a written document.

103. Find the FWD button on your phone. Just because your phone is ringing or buzzing doesn't mean you have to answer it. If you're in a meeting, speaking with someone, or otherwise engaged, plan to check the message later.

By the way Don't text and talk. Many people mistakenly think it's okay to send a text message while involved in a face-to-face conversation. Save the text for later.

104. Use good manners when balancing face-to-face conversations with phone calls. If you must check the phone for an expected or urgent call, ask the other person to pardon you as you glance at the phone. Say, "Excuse me while I check this," for example. Then make eye contact and ask to resume the conversation.

If you must answer, say something like "Please pardon me . . . I need to take this call." Be sure to step away briefly to take the call. If you anticipate a long phone conversation, ask the caller to hold, turn to the other person, and ask if you can resume the conversation later. This courtesy shows respect and will likely be reciprocated.

105. Begin and end your voicemail call with your name, affiliation, and phone number. The person listening to your message will be grateful to have this information handy without having to listen to your message repeatedly.

WEBINARS AND TELECONFERENCES

Nothing beats a one-to-one conversation. However, today's work environment mandates that we do much of our communicating by computer and phone. Since you and your audience can't always see each other to interpret nonverbal gestures, your telemessages must be organized and your communication style must be precise and focused.

106. Be an active participant, even if it's only an active listener. Many people think that if they are not a leader, no preparation is needed. However, every person on the teleconference should prepare for the meeting by going over the agenda, doing necessary research, coming up with comments, and thinking carefully ahead of time when to speak.

107. State your name before you speak. Be sure to state your name and affiliation before speaking. For example, "This is Dr. Carter from Johns Hopkins. I'd like to ask Dr. Li a question." Obviously, if you are having an informal conference call with colleagues, you may choose not to say your last name or affiliation.

108. Offer your idea first, and then give evidence to support it. "We might be able to finance this project through another department. Maybe we could try Human Resources since the project relates to training." When you offer your idea first, you help the audience make a quick and educated decision that often lands in your favor. By contrast, when you tell a story and then get to the point, you risk losing the attention of the listeners.

109. Be concise. This is a hard tip to follow, especially if you enjoy brainstorming with colleagues. The best way to ensure your own conciseness is to study the agenda and plan your comments carefully. Remember: you have only a few seconds to capture and retain the listeners' attention.

110. Use notes. The beauty of using notes during a phone call is that no one can see you reading them. Notes help you stay focused and avoid saying "um."

111. Press "mute" when you're not speaking, but don't be tempted to multitask. Pressing "mute" helps prevent others from hearing if you blurt out irrelevant comments. If you try to multitask during a conference call or webinar, you may get caught off-guard when you're called on and won't know what the question was.

SKILL BUILDER HOW TO RUN A PRODUCTIVE WEBINAR OR TELECONFERENCE

On occasion you may need to set up and run your own teleconference. The following tips come straight from our audience members who organize webinars and teleconferences routinely.

Choose a reliable webinar service. You don't want to wonder whether people on the other end are having technical difficulties. Avoid the temptation to use free webinar services; you get what you pay for. By choosing a proven webinar service provider, you get better call quality for a large audience.

Offer the audio. Give webinar participants both a computer option and a telephone option for audio. This option is especially important if you have an interactive question-and-answer session; not everyone in the audience will have a microphone on his or her computer.

Consider your audience. When scheduling a webinar for people throughout the United States, consider an early afternoon time before people begin leaving work. If your audience is from around the world, you need to plan your webinar for a time that *maximizes* attendance with the most important people on the call—even if it means setting it up at midnight your time. Scheduling an international event can become quite complicated since morning in Asia is midnight in Europe.

Optimize registration. When someone signs up for your webinar, it's your primary chance to learn something about that person. Use the sign-up as a way to gather demographic information on each participant, so you can better gauge the audience and tailor your presentation to that particular audience.

Ensure participants can reach you offline. Be sure to give all webinar participants your cell phone number in case of an emergency. There is nothing more frustrating for participants than not being able to hear, see, or follow along during a webinar. Most will just quit if there isn't an easy helpline.

Be well organized. Gather the items you'll need for your meeting and review them ahead of time. If you anticipate that you'll need to speak on a topic, plan and practice what you'll say. This preparation will minimize those annoying "ums" and rambling thoughts.

Keep the audience engaged. Pacing is important. You can tell when a live audience is starting to tune out because you can see their faces. A webinar won't give you that same visual feedback. Unfortunately, you can't see your audience surfing or checking their email instead of paying attention. You typically start losing people's interest when one topic lasts longer than seven minutes, so switch things up to maintain interest. Run through your slides several times before the webinar to get your timing down—if any concept runs long, figure out how to subdivide it.

Be sure to send thank-you notes to any participants, including both subject-matter experts and those who worked behind the scenes following the webinar. This will ensure their participation in any future activities you might want to plan.

PRESENTATION SKILLS

What is your impression of a person who is a great public speaker? Smart? In control? Moving up? Rarely will you see the president or CEO of an organization give a shoddy presentation. Somewhere along the line, these professionals have learned to present their ideas in a way that captivates and influences their audiences.

Often we begin our presentation skills training with this question: What makes a great public speaker great? No matter who is in our audience, we usually hear the same answers. Great speakers are

- In control of their bodies, the content of their talk, and the audience
- Enthusiastic and passionate about their subjects
- Interesting to watch and hear because their delivery is smooth and the material is relevant
- Engaged with their audiences and their opinions
- Able to laugh at themselves

So how do these great speakers achieve these skills? Are they born with them? Probably not. Even the most outgoing and gregarious people can benefit from speaker training. We think great public speakers are those who strive for excellence. They seek training and feedback for their presentations. They prepare meticulously and practice ahead of time. They develop their skills and thus their reputations as great presenters. As we wrote this section, we grappled with how much information to provide. We anticipated that you would choose this book because it's a quick read and we didn't want to overload you. True to our mission to write the way we teach, we refined this section to include our best tips only. Implementing the ones you like best is up to you.

PLANNING AND PREPARING

Over the years, we have taught thousands of highly educated, bright people how to give great presentations. Many were accomplished researchers, professors, medical practitioners, and businesspeople. Yet somehow in their extensive education and training, they missed learning the skills of how to pull together a great talk. When asked to present their work, they figured, "I know this material cold. After all, it's my research/idea/topic! Certainly I can pull off a thirty-minute talk." This mentality is exactly what prohibits these smart people from looking smart when they present. Instead of gliding about the room, confident and prepared, these individuals stumble through their presentations, appearing surprised by each new slide.

We once coached a physician who retired after thirty-seven years of doing research at the National Institutes of Health to become an expert speaker for a pharmaceutical company. His credentials were impeccable; he was the rock star of his topic. The pharmaceutical company's slide deck referenced his widely published work. All he needed to do was stand up in front of the audience and talk about the slides, right? Wrong. In his first presentation, this doctor droned on and on, rambled off the topic, and bumbled through the slides. Not surprisingly, the audience crushed him in their evaluations. He wasn't nervous, he knew his material inside out, and he had a great slide deck. So what was missing? Preparation and practice. Before his next talk, we went over each slide with him and asked: "What is the most important point of this slide?" Of course, the research physician caught on quickly and began to answer the questions himself, even adding some interesting anecdotes to make his points, thereby preparing and practicing for his next talk, which was a raving success.

The tips that follow come straight from our coaching sessions. We know they work because we've tried them ourselves, and we've seen them work for others.

112. Don't procrastinate on preparing for your talk. Set aside time daily to think about and make notes for your talk. The night before your presentation you should be reviewing your talk and getting a good night's rest, not designing slides.

113. Set clear objectives for every presentation before designing any visual aids. Most people start to prepare a presentation by opening PowerPoint™ and staring at the blank slide glaring back at them. Instead of getting tangled up in designing the first slide, try creating a written list or a mind map of important ideas.

114. Know your audience. Find out everything you can about your audience ahead of time. Ask your contact person what the audience knows about your topic and what they expect to learn. Never assume that you know what they want without asking. During your talk, reference your audience research by saying something like, "I understand from Dr. Rusevlyan that you want to know . . ."

SKILL BUILDER HOW ADULTS LEARN

Even if you're not a professor, your presentations will contain information that is new to your audience, thereby making you a teacher. The key to teaching adults is to understand how they learn. Technical writer Stephen Lieb summarizes several adult learning theories in a tightly written essay that coincides with our methods.[11] Note the following strategies for teaching the adults who may be in your audience.

Adults

- **Are autonomous, self-directed, and goal-oriented.** Adults want to be in charge of their learning and often have specific goals for taking a class, attending a seminar, or listening to what you have to say. At the beginning of our seminars, we often ask our audiences, "What is your greatest communication challenge?" or "What do you expect to learn today?" In this way, we're acknowledging their desire to set the agenda of the presentation while meeting their learning needs. Asking the audience questions always yields positive results.

- **Learn based on their life experiences and existing knowledge.** Adults learn by taking new information and assimilating it with what they already know. The key to accessing this information is finding out what your audience knows before you arrive, creating appropriate examples for your presentation, and acknowledging their education and experience. For example, "As physicists, you already know about the theory of uniformly accelerating charged particles."
- **Are relevancy-oriented.** Give your audience a reason to pay attention early in your talk. We always offer an agenda that highlights both the sections of the talk and their relevance to the audience, for example, "Today we will discuss how workplace safety affects your bottom line." We also like to show relevancy with transitional words and phrases, such as "as we discussed," "by contrast," "however," "in addition," and "therefore," and summary statements, such as "now that we have covered." Please see chapter 3 for more tips on transitions.
- **Are practical.** Adults understand the value of information and the time it takes to learn it. We have all sat through boring lectures thinking, *I have so many other, more important things I could be doing now.* Make your talk worthwhile: use examples that give credence to what you're saying and involve your audience in the discussion. For example, "I know many of you have developed financial tracking software. What kinds of problems have you encountered in implementing it at your organizations?"
- **Appreciate respect.** Show respect for your audience by thanking them for their time, being prepared for your talk, and acknowledging their expertise. For example, "I want to thank you for spending your lunch break with me.

Today I'd like to facilitate a discussion of recent hospital admissions." You can also show respect by encouraging the audience to participate in the discussion and weaving their comments into your talk. For example, "Thank you for mentioning the new policy. Here's how it will affect our group."

115. Make sure you are an expert on the topic. Don't accept a speaking engagement unless you can fulfill the audience's needs and expectations. Audiences are quick to recognize posers.

116. Don't try to tell your audience everything you know. *This is one of the most important tips in the book.* Your goal isn't to impress your audience with how much you know. Rather, your goal is to impress them with how much you know about them and to be able to tell them exactly what they need to know. Choose three or four key points to support your purpose and objectives and offer relevant examples only. Creating an agenda will help you plan your talk. Give the most important information only, offer specific examples, and use the power of storytelling to illustrate your points. Remember what Voltaire said: "The secret to being a bore is to tell everything."

117. Plan to keep it tight and focused. Make sure every one of your stories, examples, and slides is relevant to the audience and your purpose. If you begin to ramble on about other topics or need to give detailed explanations for your slides, you'll instantly lose your audience.

118. Prepare with enthusiasm. People often describe great speakers as having enthusiasm that is "contagious." While it doesn't seem logical that you could learn to be enthusiastic, in our speaker coaching experiences, we have seen that you can. Enthusiasm is all attitude. If you plan and prepare to do well, look competent, and sound articulate, you'll appear enthusiastic.

SKILL BUILDER HOW TO DESIGN THE OPTIMAL SLIDE

Creating a slide deck to accompany your presentation can help you organize your thoughts, plan your examples, and consider your audience's needs. The key to creating great slides is allowing enough time to put them together in the most appropriate format.[12]

Keep the design horizontal with a 2:3 ratio. This eliminates the need to adjust the display screen that shows your slides.

Limit content to one subject. Each slide should make only one point. If you're comparing data and have several items to show, make sure that the data on your slide illustrate your main point. You don't want your audience to have to search for the evidence. Don't crowd the slide with multiple graphics, even if they convey the same point.

Use a maximum of five to seven lines of text, including the title. This is where you can take advantage of bullets. Look carefully at your points and write them with parallel structure, in bulleted form. Please refer to chapter 3 for tips on writing with bullets.

Use no more than seven words per line. Here's where you'll need to write with density by choosing action verbs and using fewer words to illustrate your point. Remember: slides are a visual aid, not a transcript of your presentation.

Choose an optimal, sans serif font such as Calibri. Fonts without "feet" are easier to read when projected. Likewise, fonts with "feet" (serif fonts) such as Times New Roman are easier to read in print.

Use point sizes that are large enough for everyone in the audience to read: 44 point for headlines and 32 point for body copy.

Keep your slides simple. You should not need to apologize for the quality of your slides. It's an insult to show your audience a slide that is hard to read or confusing; it implies that they weren't important enough for you to offer your best work. Here are additional, more specific tips for making your slides easier to read.

- Label parts of the graphic directly on the chart rather than to the side in a key or legend.
- Limit bar graphs to 5–7 bars (vertical design: 4 columns; horizontal design: 7 rows).
- Create pie charts with 5 or fewer slices; put percentages inside the pie pieces and labels nearby.
- Design line graphs with a maximum of 3–4 lines.
- Avoid using ALL CAPITALS and *italics* as both can be harder to read than plain text.
- Keep color choices simple. Most graphics are easier to

comprehend with variations or gradients of one color rather than different colors.[13]

Avoid unnecessary slide features. Avoid trying out trendy colors, sounds, unrelated graphics, and any other unnecessary but exciting features as they distract the audience and impair learning.[14] Your slides should enhance what you say, not draw attention from you and your content.

Choose pleasing color combinations: a blue background, yellow headlines, and white body copy, or a simple white background with black type. While it's tempting to use unusual and different colors, this tactic may backfire as your audience focuses on your slide design and not your presentation. Remember: most color-blind people typically can't distinguish between red and green so trying to emphasize data with the color red isn't effective with them.[15]

By the way If you like to mix and match your slides for different talks, make sure you always use the same design template. That way you can grab slides from another deck without having to redesign them.

119. Plan to arrive early. You can fix lots of potential problems (such as poor room setup and equipment failure) if you arrive early enough to deal with them. Another perk of arriving early: you can make friends with the professionals who can help you (audio visual technicians and wait staff).

By the way Make time to greet your audience members as they enter. Working the room will give you a feel for what your audience wants and will help you gain their support. By the time you begin to speak, you'll have a relationship with them.

ORGANIZING YOUR TALK

Common sense and adult learning strategies dictate that your talk should be organized in a logical manner. We suggest the following sections.

THE INTRODUCTION

The introduction is the most important part of your talk. Give it the attention it deserves.

120. Plan and practice your introduction. Many presenters struggle with the first part of their talks. These speakers may be a little nervous at first and need to get going to smooth out their delivery. We have heard many speakers begin their presentations with something like this: "Um, uh, I'm Peter Jones, and uh, I'm here to talk about, uh." Avoid this stuttering with a well-rehearsed introduction. Our trick: write down the first few words you plan to say and glance at them right before you stand up to speak.

121. Avoid the "running start" introduction. This is an introduction full of information the audience already knows. Generally, presenters use the "running start" to warm up before getting to the "real" stuff. For instance, a pulmonologist might start a talk to fellow pulmonologists, "Cigarette smoking is the most common cause of chronic obstructive pulmonary disease." Clearly information any pulmonologist already knows.

122. Avoid "data dumping" in the introduction. Don't fill your introduction with statistics and numbers that no one in the audience will ever remember. For instance, "American young adults (ages 18–21) earned about $211 billion in 2003. This group is spending at a rate of approximately $172 billion per year; and in 2005, savings rates dipped to minus 0.5 percent."

By the way Both the "running start" and the "data dumping" introductions are often used as a way to cope with nervousness or a lack of preparation. Check your introduction; don't make these common mistakes.

123. Start your presentation with something worthy. Your audience decides whether or not to pay attention during the first thirty seconds of your talk. Draw them into your topic. Two ways to accomplish this:

- **Tell a story.** For example:
"I once had a colleague . . ." Make sure your story is tight and relevant by rehearsing it.
- **Ask a question.** For example:
"Do you ever struggle with the overtime schedule?"

124. Tell your audience your purpose. Appeal to their adult learning sensibilities by telling them why you're there. For example, "I'm here to share my research on the environmental problems that pose a risk to public health in North America. My goal is that when you leave today, you'll . . ." (Here's a great place to put an action or benefit statement.)

125. Establish credibility in your introduction. After you hook the audience, introduce yourself by telling the audience something about you that gives you credibility. For example: "As you know, I'm Callie Bennett. I'm very interested in genetic testing. In fact . . ."

By the way You should offer the credibility statement even if you have been introduced by someone else. You don't need to review all of your credentials; simply mention what is relevant to your topic and your audience.

126. Tell your audience what is coming with an agenda. After you establish credibility and give your purpose, offer the agenda. For example: "Today we're going to cover three areas."

Offering an agenda seems so elementary. We've all heard the presentation advice, "First tell them what you're going to tell them. . . ." Doesn't everyone have an agenda? In fact, they don't. One of the best ways to get started on a talk is to set an agenda and build your content and examples around it. Remember: the agenda is even more important in a talk that doesn't have slides.

THE BODY

The body of your talk should include your main points and the examples to support them. The best way to set up the body is, depending on your time, to choose three to five main points and discuss *only* those. As you select each point and the example to illustrate it, refer to your purpose statement and ask yourself:

- Does this point support my purpose?
- Is this story/example relevant to my audience?
- What do I want my audience to do with the information I offer?

127. Use transitions to guide the audience through your points. For example, after you offer the agenda, make a transition statement to lead your audience into your first point: "First let's talk about the chemical basis of . . ."

128. Choose examples and wording that suits your audience. For example, "For those of you who . . ." People often ask us, "What do you do if your audience has different levels of expertise?" We offer two recommendations:

■ **Acknowledge the varying levels.**

For example:

"I know some of you are new to the electron microscope and others are experts . . ." We also suggest asking the audience a question: "How many of you have ten or more years of experience? Five years? One year or less?" In this way, you can refer to people in the audience when you're making points that call for certain levels of knowledge. Another way to address different levels of expertise is to have specific examples ready: "What this means to you as nurse practitioners is . . ." and "What this means to you as pharmacists is . . ."

■ **Clarify your examples by qualifying them.**

For example:

"When we designed this building, we paid special attention to the seismic resistance, or its ability to withstand earthquakes." As long as your explanation is brief, you'll not bore the people in your audience who already understand it. You may also preface your comments with "As you may know" or "As many of you know." Notice we don't suggest saying, "As you know" since everyone may *not* know.

129. Use visual aids to enhance your presentation, not to replace you. Don't make the mistake of assuming that if you have slides, you have a presentation. Ideally, you should be able to give your talk without your slides (or other visual aids). Slides should not be a teleprompter. Rather, slides should be a way to reinforce your points.

130. Don't read your slides aloud! Reading your slides to the audience isn't necessary, as most people can read silently three times faster than you can read to them.[16] Plus, why should your audience listen to your talk if they can simply read the slides?

(It's okay to read a headline or bullet point, and then talk about it in your own words.)

131. Control the laser pointer! Nothing highlights a nervous twitch like a laser pointer. If you need to point out something on your slide, use the following techniques.

- **Shoot from the hip.** Prop your hand on your hip to help steady the flicker of light.
- **Point and describe.** Use verbal cues to accompany the visual ones. For example, while pointing at the third column, say, "In the third column you can see . . ."
- **Keep it under control.** Once you have highlighted the specific area of your slide, put the laser pointer down.

By the way If your slides are thoughtfully designed, you may not need a laser pointer to highlight important data. There is nothing more distracting than a bouncing light on every word of a slide or an out-of-control beam swirling around the room.

132. Don't depend on your slides to tell your story. We often have people ask us, "How many slides should I have for a thirty-minute talk?" There is no rule on how many slides to have. It depends on your audience, purpose, and topic. As you plan, design, and rehearse with your slides, ask yourself, does this slide

- Add value to my talk?
- Meet the needs of my audience?
- Serve as a teleprompter for me or as a visual aid for my audience?

THE CONCLUSION

Often, presenters find themselves racing through their conclusions because they don't budget their time wisely. Make sure you give your conclusion the emphasis it deserves. Why? Because it's often the only part the audience will remember. Slow down and make the conclusion memorable. The conclusion of your talk should be a summary of your main points, not a repeat of your entire presentation. This is your chance to give credence to your argument by weaving everything together in a neat package. Here's how.

133. End your presentation with a take-home message. The conclusion is your opportunity to recap important points to help your audience retain or take action on them. Two different ways to accomplish this:

■ **Offer a basic summary with an action statement.**
For example:
"Today we talked about the role that water plays in regulating the global environment. Three things I want you to remember about water are . . ." This action statement should refer back to your original purpose statement.

■ **Give the audience a mini-quiz.**
For example:
"Let's do a quick review. Can someone tell me what three things influence precipitation measurements?" The mini-quiz is a perfect choice to end a talk when you're running out of time. Don't think of it as patronizing your audience; they will be glad to show their understanding, especially if it helps conclude your presentation.

By the way Be sure to thank the person who invited you to present and the audience for their time. This thank-you will go a long way to help you meet your future goals.

ANSWERING QUESTIONS

If you master the art of answering questions, especially the ones that interrupt your presentation, you'll gain the attention and respect of your audience. Here's how.

134. Help the audience develop questions by providing direction. For example, "Are there any questions about the Clark Device?"

135. Give your audience time to digest what you've said. If you ask a question, make sure you give the audience ample time to answer. Typically, large audiences (more than fifty people) will not answer. That's okay. Give them a few seconds (count silently to three) to think of the answer, and then offer the answer as a transition to your next point. If there are still no questions after several seconds of silence, you can prompt the audience with a sample question, for example, "One question that often comes up . . ."

136. Repeat or paraphrase the question in your answer. This tip is hard to remember yet crucial to the presentation. If the audience can't hear the question, they will not pay attention to the answer.

For example:

> "I'm glad you asked that. What is the best research method? (Pause for three seconds.) I prefer . . ."

By the way If you're presenting to a large group (more than fifty people) or speak softly, ask for a cordless microphone. Let that mic on your lapel be your reminder to repeat all questions.

137. Be honest when you don't know the answer. It's okay to defer to the audience for the answer. For example, "I'm not sure about that. Does anyone have experience with this situation?" Use your open palm to gesture to the audience when you ask this question.

138. Focus on sharing ideas and information rather than on giving advice. The best presenters take advantage of the audience's participation. When the audience contributes, they feel invested in the content, you appear to be a collaborator instead of a lecturer, and they do some of your work for you!

139. Use these techniques for difficult questions.

- **Redirect the question to another person in the group.** If you know that someone in your audience may have the answer, ask him or her to comment. Of course, never put a colleague "on the spot" to answer a question unless you know it's okay. If you arrive early, as we already suggested, you can ask this expert if it's acceptable to defer some questions to him or her.
- **Probe or clarify the question before responding.** For example, "Let me make sure I understand the question. You want to know . . ." This technique will buy you time while you formulate an answer.
- **Postpone a response if the question isn't relevant or is too complex.** Your audience will love you if you stay on task and avoid wasting their time with unrelated details. For example, "Thank you for mentioning cost-cutting strategies, Frankie.

Today's presentation is geared toward recruitment. I know others may want to hear more on cost cutting later. Do you mind staying afterward to discuss it?"

▪ **Refer to a visual aid.** Nothing says, "I'm prepared" better than "I have the answer to that question on my next slide."

140. Don't let your audience control the discussion. Some people will ask questions or make comments just to hear themselves talk. This is okay as long as

▪ Their contribution is relevant.

▪ You maintain control of the floor. Do this by summarizing what the person said and making a transition to your next point. For example: "You're right, current technology does not support this method. Let's take a look at what does."

By the way Occasionally you may have people who continually interrupt the presentation with comments or questions. Two ways to stop a disruptive audience member from speaking up:

▪ Break eye contact and don't make eye contact with him or her again.

▪ Stand behind him or her (if possible).

QUALITIES OF EXCELLENT SPEAKERS

When he was alive, Mark Twain earned a meager living as a writer. His primary source of income later in life came from presenting.[17] He traveled all over the world performing the way stand-up comedians do today. The secret to his success? He was an excellent storyteller.

141. Tell stories to illustrate your points. The best public speakers are great storytellers. As we discussed earlier in this chapter, most adults learn by taking new information and assimilating it with what they already know. When you tell a story, you give your audience a prompt to locate what they already know so they have a place to "put" your story in their knowledge bank. In the words of one learning expert, "stories enable us to engage with new knowledge, broader perspectives, and expanded possibilities because we encounter them in the familiar theme of human experience."[18] Adult learners form opinions about your topic based on how well you relate it to them.

When we suggest that you tell stories to illustrate your points, we don't mean that you should recite a novella. Your stories should be tightly crafted and well rehearsed and contain brief examples. Some of the best "stories" are thirty-second analogies or questions that lead the audience into the topic. For example, "Did you know that your garden shed probably contains all the ingredients needed to make a small bomb?" Notice how this example plays into everyday life yet alludes to serious chemistry.

The places in your presentation where you should tell a story:

- At the very beginning of your talk to capture the audience's attention
- After each point to illustrate it
- At the end of your presentation for a summary

> *By the way* For your stories to be effective, you must practice them. Nothing is more boring than the speaker who tries to illustrate a point with an unrehearsed mini-dissertation. No time to practice? You don't have to rehearse your entire talk at

once, and you can rehearse "in your head" instead of out loud. Try practicing bits of it while driving to work, taking a shower, or exercising.

142. Plan and practice ways to appear enthusiastic. Two ways to do this:

- **Use eye contact and facial expressions to connect with the audience.** When you make an important point, make sure you're looking directly at the audience. Pause, smile, lean your head slightly toward the audience, and make sure your eyes are open wide. Try to make eye contact with each person in the audience at least once.

  By the way If you're presenting to a sea of people and the audience is too large and far away for you to see individuals, divide the group into quadrants and make eye contact with each section as you present. The people in each group will perceive you're looking directly at them.

- **Gesture with your hands to illustrate your points.** One of the best ways to appear enthusiastic about your content is to gesture. The key to gesturing is to keep your hands out of your pockets, off your hips, unclasped, and in front, where they are free to gesture. The easiest way to do this is with the "L" Shape Ready-to-Gesture Position. Bend your arms at your sides, so that they make an "L." It's okay to have your fingers touch lightly at the tips between gestures. At first, gesturing may seem a bit awkward. Do it anyway. We often videotape speakers to show them how they look and sound when they present. The biggest surprise to them is how their planned gestures don't look staged.

By the way Always gesture to your audience by offering an open palm hand rather than a pointed finger.

143. Position yourself where everyone in the audience can see you. At the beginning of your presentation, choose a spot in front of the audience where everyone can see and hear you. Mark this spot mentally with an invisible X (or with a coin, if necessary).

144. Move with a purpose. Stand on the invisible X throughout your presentation. When you have an important point, a question from the audience, or a new example, step forward (off the invisible X) to emphasize it. Always return to the invisible X so your audience will not have to shift positions to see you. (Returning to the invisible X is especially important when the room setup consists of round tables and the participants must move their chairs to see you.)

When you want to list a few important points, try the following movements. Step to the side with your arm outstretched and say, "My first point," then sidestep again for your second point, and so on. Remember, speakers who move without a purpose (such as walking around too much or bobbing back and forth) appear nervous and unprepared.

145. Use the lectern to your advantage. Oftentimes the room where you'll present will have a lectern that you must use. The lectern is a great place to put your notes and equipment. It may also have a microphone that you need to use. If you stand behind the lectern, be sure to take your hands off it. You'll naturally show more body language. In addition, be sure to gesture, vary your facial expressions, and enunciate.

If at all possible, ditch the lectern altogether. Your audience will enjoy being able to see and interact with you, and you'll be able to use your whole body to gesture. The disadvantages of using the lectern are that it

- Acts as a physical barrier between you and the audience
- Is usually positioned near the front of the room, creating greater distance between you and the audience

146. Use voice inflection to emphasize points. As you plan your presentation, make a note of the most important information and make sure you either raise or lower your voice when you say it. Avoid using qualifiers. Some qualifiers that don't add value to your point are "always," "quite," "really," "somewhat," and "very."

For example:

Instead of
"In this situation, I *really think* that we should try a new processor."
Say
"In this situation, I *recommend* that we try a new processor."

Deleting the word "really" and using the word "recommend" adds credibility to the comment. You should place the emphasis on the end of the sentence, where you state your recommendation, since it's the most important point. You can emphasize either "new" or "processor" with voice inflection, depending on what you want to emphasize.

> *By the way* A common problem we see when coaching speakers is inappropriate voice inflection. Instead of emphasizing

important words, some speakers either drop off or speak up at the end of phrases and sentences. This habit distracts the listener who is waiting for a verbal cue to find out what is important. Try this exercise: Record then listen to yourself reading the following passage aloud, emphasizing the words in bold:

> I have just completed a **study** on the effects of plant **health** on food **safety.** In this **study** we looked at the **outbreak** of **Salmonella** on tomatoes, seed sprouts, cantaloupe, and **apples.**

Did you notice how the emphasis of words at the end of phrases actually hindered comprehension? If you allowed your voice to lower, or trail off instead of lift up, you would have a similar result. Your audience will wonder, "What was that?"

Now try reading the passage with only important words emphasized.

> I have just completed a study on the effects of **plant health** on **food safety.** In **this study** we looked at the outbreak of **Salmonella** on **tomatoes, seed sprouts, cantaloupe,** and **apples.**

147. Vary the speed of your delivery. Slow down for important points and speed up slightly at less important points.

By the way The best way to improve your presentation skills is to watch yourself. Videotape yourself practicing your talk so you can fine-tune your delivery.

148. Learn to think quickly and logically when you present. The most dynamic speakers seem to thrive on audience participation, even when it causes the discussion to deviate temporarily from their planned presentations. This skill is the direct result of preparation and practice. When you have command of your content, it's easy to step out of sequence briefly to address audience questions and comments. Speakers who think quickly and logically are willing and able to change content on demand.

149. Communicate clearly and expressively. Great speakers know that to communicate clearly, they must enunciate, use voice inflection, and choose words carefully. One easy way to do this: open your mouth wider. This tip may sound a bit strange, but it really helps prevent mumbling.

150. Slow down. One of the most common problems we address in coaching speakers is pace. Theytalkreallyfastbecause theyhavealottosay. It sounds a lot like that last sentence looks: jumbled.

151. Use verbal cues to direct your audience. Another important way to communicate clearly is with word choice. Since your audience can't reread what you say, you should offer them verbal cues to identify what is important and where you're going.

Use verbal cues to

- **Highlight important information.** "The most important point . . ."
- **Make transitions between points.** "Now that we have talked about constructing your portfolio, let's talk about how you can minimize your tax burden." (Notice how the "you/your" focus sets up a benefit statement.)
- **Repeat key phrases.** "Remember when I mentioned this example earlier? Here's how it relates to you."

152. Avoid trendy phrases. In an effort to relate to the audience, many speakers are tempted to use trendy phrases. Some examples include "at day's end," "if you will," and "out of the box." (See chapter 3 for examples commonly used in writing.) The problem with these phrases is that they really don't add value to the content or delivery. In fact, many speakers are mocked for their overuse of these phrases, such as the regional director we recently heard tell his team, "At the end of the day, moving forward, we'll need to huddle up and encourage more 'idea showers' and out-of-the-box thinking if we want to take it to the next level." We recommend steering clear of these expressions or words, if for no other reason than to keep your audience from using you as material for their next comedy routine.

What to do instead?

Use the public speaking tips we have already mentioned:

- **Pause.** It's much more dramatic.
- **Gesture.** Your audience will listen more attentively.
- **Practice.** Many people use trendy phrases as fillers when they lose their train of thought.

153. Work with a coach on pronunciation. If English is a second language for you, ask a friend or colleague to help you with pronunciation and usage before you present. If you as an English speaker are presenting to a non-English-speaking group, you may want to practice pronouncing common words in the audience's language. For example, if you are in presenting in China or India, you should learn how to say "hello," "goodbye," "please," and "thank you" in their respective languages.

154. Engage your audience with active listening skills. Active listening allows you to connect to your audience. As we said earlier, great speakers make a point of appearing enthusiastic and engaged. You can engage others by using eye contact, listening attentively, asking probing questions, and paraphrasing audience comments.

155. If you want to probe for more information, ask, "Has anyone ever experienced this?" Make sure probing for more information doesn't take you off topic and waste time.

156. Leverage the expertise of others. If you're giving a talk and know that an expert on the subject is going to be in the audience, we often suggest you try to contact that person ahead of time to discuss the topic. That way, you can let the person know that you're interested in hearing his or her opinions. You may also ask if it would be all right to defer to him or her for any difficult questions. In this way, you turn a potential opponent into a support system.

157. Convey warmth to others. How do you convey warmth? This sounds a bit nebulous. Some of the communication tips we have covered can help you. First, sincerely care about your topic. Then use these nonverbal cues: make eye contact, smile, and gesture. It's also nice to be a little self-effacing and have a sense of humor.

158. Be able to laugh at yourself. There is a difference between taking your work seriously, which is imperative, and taking yourself too seriously, which can be disastrous. Being able to laugh at yourself will help you relax in front of your audience, look in control, and appear approachable.

159. Use humor wisely. People often ask us if they should use humor in their presentations. We think you should use humor only if your joke is

- Funny (and your delivery is well rehearsed)
- Appropriate and relevant
- Not offensive to the audience

160. Demonstrate self-confidence. We can't give you a formula for self-confidence, but we can tell you how it appears to others. Self-confident speakers are well prepared. They don't mumble or speak in a monotone voice; they make eye contact with each member of the audience; and they appear to enjoy the challenge of interacting with the group. In sum: they are prepared, they respect their audiences, they are enthusiastic, and they practice.

Never make the mistake of thinking, "I know this so well I don't need to practice or prepare."

161. Show leadership. Leadership is another one of those terms that has prompted businesspeople to write entire books. We think that leadership is fundamentally based on the Golden Rule: Do unto others as you would have them do unto you. In presenting, this translates into showing others how a great speaker gives a great presentation. This concept goes back to the original question at the beginning of this section: What is your impression of a great speaker? When we ask our audiences this question, nobody ever says "ill prepared."

162. Have a business orientation. If a group of busy people has gathered to hear you speak, make it worth their time. Show respect for your audience's time, attention, and opinions by

- Starting and finishing on time
- Keeping the hecklers under control
- Understanding the group's needs and giving them only the most relevant information
- Allowing the audience to comment and ask questions, preferably during your presentation

163. Make sure your content is both timely and timeless. Timely content is sensitive to current events, trends in the industry, or recent changes in the organization. You can show timeliness in your content by commenting on a current issue. For example, "I know everyone is worried about the company merger." Timeless content refers to things that withstand the test of time. You can incorporate timelessness into your content by mentioning an ongoing issue. For example, "We always see a decline in sales at the beginning of the first quarter."

164. Pause for emphasis. Take advantage of the power of the pause. In public speaking, timing is everything. Make sure you plan to pause after every key point. If necessary, write PAUSE in your notes to remind you.

165. Practice. Practice. Practice. Need we say more? Of all the great tips we have offered in this book, practicing is perhaps the most important one. People in our audiences often suggest that it's possible to overpractice. They claim that too much practicing makes a talk appear too staged. We have found that the "stiff" presenters are the ones who haven't practiced. They are so busy trying to remember what they are going to say, they can't tune in to the audience or deviate from their slides. In contrast, the speakers who have mastered their content seem to glide about the room, exuding just the right amount of enthusiasm.

166. Embrace your own style. All of our suggestions on presentation skills are just that: suggestions. You should try out these tips and incorporate the ones that work best for you. Remember, your public speaking style will probably be different from your colleagues'. That's okay.

5 Meet

For many people, an upcoming meeting is simply a deadline to complete a piece of a project such as gathering data, making phone calls, and doing research. Leaders in science, technology, and medicine understand that preparing for a meeting is a great time for critical thinking and attending a meeting is a great time to collaborate, negotiate, and build support for projects.

Have you ever noticed that the people who have the most success leading others appear extremely well organized? This is no coincidence. People who manage big projects understand the value of time and money, especially when the time and money are someone else's. The head of a drug utilization review board, a major integrated circuit design project, or an emerging software company can't afford to arrive at a meeting without thoughtful preparation. Preparation is crucial to appear competent in all types of meetings: when you meet with colleagues, meet new business contacts, and meet the media.

MEETING WITH COLLEAGUES

Harvard Business Review OnPoint reports 78 percent of surveyed executives said they had slept through a presentation in the last month.[1] It appears executives aren't the only ones tired of meetings. Microsoft Corporation, makers of PowerPoint™ presentation software, offers some of the most insightful data about workers' views of productivity in meetings. Microsoft surveyed 38,000 people in 200 countries and found that workers spend an average of 5.6 hours each week in meetings and that 69 percent of these workers think their meetings are unproductive.[2] So how can we make meetings more productive? The answer is better preparation and better communication.

Thoughtful preparation and planning are essential to a successful meeting—whether your meeting is one hour or three days. In addition, effective communication skills are necessary for a fruitful meeting. In fact, meetings encompass almost every kind of communication: presenting complex data; writing agendas, summaries, and proposals; persuading others to adopt an idea; dealing with upset or difficult people; and managing time, people, and projects. Whether you're responsible for running a meeting, presenting at a meeting, or just attending a meeting, the following tips will help you stay focused and appear thoughtful and articulate.

SKILL BUILDER HOW TO RUN AN EFFECTIVE MEETING

The core problem with unproductive meetings is lack of clear objectives and leadership. If you're assigned the important task of chairing a committee or running a meeting, make sure you clearly communicate your objectives to the group. Remember, it's your job to encourage collaboration and make sure the committee members complete their assignments between meetings. Communicate clearly so you don't have to micromanage.

Set an agenda. Draft a brief agenda with topics you need to discuss, including specific objectives and allotted time. For example, under the heading "Discuss Potential Formulary Changes," you may want to list specific options.

Ask participants to arrive on time and to be prepared. To increase attendance, schedule recurring meetings on the same day and at the same time, and remind participants a few days beforehand.

Start and finish on time. Staying on time is a difficult task. Once you and your group become adjusted to starting and finishing on time, you'll notice fewer late arrivals and better-prepared attendees.

Keep the group on task. If you're running the meeting, it's your responsibility to keep things moving. A few tricks to stay on task:

- Refer to the agenda. "We need to finish this discussion so we can go on to item number three."
- Make transitions. "Lars is going to follow up with an equipment study next month. Thanks, Lars. Now let's talk about our budget."

- Summarize. "So we're going to go forth with the project."
- Bring to a vote. "May I have a motion to vote?"
- Table or park ideas. "Let's put this on next month's agenda when we have more time to discuss it."
- Intervene. "Let's stop this discussion for a moment and look at what's happening here."

Praise and thank participants for their contributions. Everybody likes to be acknowledged and thanked. This easy public gesture will help motivate your group.

Encourage participation. A few ways to foster collaboration and encourage participants to speak up:

- Ask for comments with open-ended questions: "What do you think about this idea?"
- Pause between points so participants will have time to gather their thoughts and make comments.
- Look up from your notes and make eye contact to show an open forum.

Follow up with an action plan. If at all possible, avoid documenting the meeting with passive voice minutes, for example, "It was decided." Instead, write what needs to be done next in the form of an action plan: "For the next meeting, Mai will report on the Hildebrand Project." An action plan will help your group define and complete its objectives before the next meeting by creating a record of accountability. See chapter 3 for more tips on writing in the active voice.

PRESENTING AT A SMALL GROUP MEETING

Presenting for five minutes at a board meeting requires as much preparation as presenting a thirty-minute speech at a conference. Why? It takes more preparation to say less. Most people who ramble when they present haven't prepared adequately. Refer to the presentation tips in chapter 4 and add these meeting-specific tips to them.

167. Prepare your comments in advance and practice delivering them. Nothing dissuades an audience more than an ill-prepared presenter. If you anticipate lots of discussion on your idea, make sure you're prepared to respond to questions and concerns. If you don't attend the group's meetings on a regular basis, ask your contact for insight on the group's opinions and meeting protocol. Make sure you understand what motivates the group before you try to persuade them.

168. Stand up to speak. Standing up allows you to achieve better eye contact, the key to power in a meeting. If you think standing will intimidate your audience or impede your approach, remain seated.

169. Speak clearly and loudly. Participants who speak the loudest often get the most attention and time at meetings. If you're soft-spoken, ask for time on the agenda. If you have no trouble speaking loudly, make sure you avoid interrupting others and dominating the discussion, as this will lessen the power of what you say.

170. Strive to make eye contact with everyone in the group. Your comments will have greater impact with a powerful delivery. As we said earlier, eye contact is your strongest tool. To achieve maximum eye contact.

■ Sit in the seat that offers the most eye contact, especially if you are the chair. At a long table, sit at the end or corner.

■ Sit next to the people you expect to disagree with you. They will have a harder time doing so if they can't look at you directly.

Here's an interesting eye contact story. A speaker was on her way to give a presentation at a teaching hospital when the sponsor, a pharmaceutical division manager, stopped to check on a dinner meeting she was organizing at a local restaurant. The speaker accompanied the division manager inside, a hostess escorted them to a booth, and they sat across from each other.

The restaurant's banquet manager walked up to the booth for the meeting and sat next to the division manager, across from the speaker. From that point on, the banquet manager directed all of her comments to the speaker instead of to the division manager. The division manager was the customer—yet the seating placement directed all eyes onto the person with the least decision-making power, the guest speaker. The banquet manager kept looking at the speaker for approval and the speaker kept referring her to the division manager. In this scenario, the banquet manager should have asked to sit across from the division manager to achieve greater eye contact.

171. Offer your handout after you explain your idea. As soon as you give your audience a handout, they will look at it instead of at you. Use your handout to point out details that you don't have time to cover in your allotted time slot. Exception: when you need to refer to a graphic and don't have slides to show it, offer a handout instead.

172. Give your point first, and then offer supporting evidence. We have already mentioned this idea in the writing chapter. Why say it again here? It's one of the most crucial things to do when presenting at a meeting. When you give your point first, you set your audience up to listen to your great argument.

For example:

> "I think we should try Brand X. It's less expensive, has a great safety record, and is easy to use." Notice how this statement uses benefits to persuade the group. Refer to chapter 3 for more tips on creating benefit statements.

173. Make one point at a time. If you offer two great arguments and one mediocre argument to support your idea, the group will ignore your strong ones and attack the weak one. Offer strong convincing facts to persuade others to adopt your ideas.

174. Anticipate and prepare for your colleagues' attitudes and positions. Gain the approval of the group by anticipating what others will say before the meeting. Prepare for difficult questions by planning and practicing your answers. Nothing sounds more professional than a speaker who says, "I'm glad you asked about the global implications. As a matter of fact, I can answer your question with my next slide/handout/example." An easy way to find out what the group will think of your idea ahead of time: ask.

By the way If you have people in your group who always ask difficult questions or who want to comment on every point, try speaking with these participants before the meeting. Giving them a forum to express themselves could save time, garner support for your idea, and help gain consensus in the meeting.

175. Persuade others by focusing your ideas on benefits rather than features. Two ways to add benefit statements to your ideas are

- Summarize your statement with, "What this means to you/ our organization/ the department is . . ."
- Use if/then statements such as, "If we adopt this idea, then we can save over $40,000 in our department next year."

By the way When persuading others, never reveal that you're nervous or ill prepared. Act as if you know what you're doing and others will think you do.

176. Prepare visual aids. The best visual aid is an articulate speaker. Choose other visual aids that suit the audience, room, and presentation. Make sure you arrive early enough to test your equipment before the meeting.

PARTICIPATING IN A SMALL GROUP MEETING

We have already established that most people dislike meetings because they find them unproductive. This reasoning is the basis for the following tips on how you can establish yourself as a productive team or committee member who facilitates collaboration.

177. Come to every meeting prepared. Don't waste the other members' time by showing up unprepared.

178. Make notes during the meeting and use them when you speak. It's better to refer to notes than to stumble through your message. To avoid paper shuffling and reading to your audience, make sure your notes are brief.

179. Speak only when your contribution is relevant. Speaking up often on irrelevant matters diminishes the power of the important points you wish to convey.

180. Speak the language of participants. Using the terminology of your peers will show them your ideas are thoughtful and well prepared. As we said before, avoid using jargon.

181. Be specific and offer evidence to support your points. Any claim you make should contain valid evidence to support it, preferably from the organization.

For example:

"I checked with our human resources department, and they said we do have a hiring freeze right now."

182. Respect people's time by presenting materials simply. The biggest complaint people have about meetings is that they last too long. For this reason, presenting your ideas in a simple, concise fashion will give you the advantage of appearing focused and prepared. Remember, never compromise content for simplicity.

183. Show respect for other people's ideas. The best way to gain respect for your ideas is to listen to the ideas of others. Demonstrate this courtesy by paraphrasing. When you echo what others have said before you speak, you demonstrate that you have been listening and that your comments are thoughtful. Once you rephrase what others have said, avoid connecting it to your comments with "but." Connecting the phrases together with "but" negates what the other person has said and can make you appear patronizing. Instead, choose "and" or just end the sentence.

For example:

Instead of
"I agree with Tom that we should consider replacing our fleet with new vehicles, *but* I think we should check into the new hybrids before we make a final decision."
Say
"I agree with Tom that we should consider replacing our fleet with new vehicles. I *also* think we should check into the new hybrids before we make a final decision."

In the previous example, offering your idea at the end of the sentence places it in the stress position, the more persuasive location. For more information on the stress position of a sentence, refer to chapter 3.

184. Monitor nonverbal signals. Your colleagues and your boss will see your behavior at meetings. Make sure you appear professional.

- Avoid slouching, looking away, and doodling on your note pad.
- Turn off potentially distracting items such as cellular telephones, laptop computers, and handheld communication devices. Often we have seen people busily texting and checking messages during meetings so as to appear indispensable and important. Instead, these people appear both rude and inattentive. Communication professor Dr. Clifford Nass and his colleagues at Stanford University note that the more people try to multitask, such as texting during a meeting, the worse they are at participating.[3]

MEETING NEW BUSINESS CONTACTS

Let's say you're traveling alone to an industry conference that opens with a reception. The idea of pasting on a smile and making small talk with a bunch of people you don't know makes you queasy. But you have to go. What to do? For some people, the idea of entering a roomful of strangers and creating conversation is worse than having a root canal. Yet we all know that many business relationships begin in these networking settings.

Start with these steps for entering a room with confidence.

Enter with a smile. Nothing makes others feel more welcome than a smile. You'll appear approachable and you'll feel more confident.

Go get a drink, preferably nonalcoholic. Your drink will give you something to do with your hands and an excuse to break away. "Please excuse me, I need a refill."

Introduce yourself. This is easier than it sounds: "Hi, I'm Noah Robinson from ABC Hospital."

Say something nice about the other person. Compliments will always open doors to conversation: "I enjoyed your talk today," or "I love that scarf."

185. Offer a firm handshake. It's okay for both men and women to offer a handshake in business. Make sure it's firm but not overbearing. Use this moment to make eye contact, smile, and repeat the other person's name so you can recall it later.

By the way Place your nametag on your right side rather than your left. Most people follow their right-hand shake visually and will naturally look at your right shoulder to see your name badge.

186. Say, "It's nice to *see* you" instead of "It's nice to *meet* you." Using the word *see* saves you from the embarrassing response of "We've already met!" from the other person.

By the way If you have forgotten someone's name, introduce yourself. This will prompt the other person to say his or her name.

187. Make sure you can describe your work, career, education, employer, and current projects in a nutshell. Nothing bores other people more than a lengthy explanation to a simple question. You don't need to be overly simplistic, just concise.

For example:

"I'm a mechanical engineer for XYZ Company. I design containers that house our petroleum products." Obviously, when you're meeting peers in the industry, your explanation should be more specific and technical, for example, "I'm a senior mechanical engineer for XYZ Company. I oversee the tank systems in the metro Chicago area."

188. Learn to be a great conversationalist. Being a great conversationalist requires a balance of talking and listening. For most of us, it's much more fun to be the storyteller than the listener. Fight the urge to talk about yourself and your projects the entire time. Mix in a few questions for the other person(s) in the group and be attentive.

189. Be humble. If you're outgoing and enthusiastic about your work, this tip may be as hard for you as it is for us. Remember that although you may have achieved a lot, such as a successful career, an excellent education, a big grant, or a great promotion, you didn't get there by yourself. Someone else taught, nurtured, or recognized you and therefore contributed to your success.

190. Assume other people know about you before they meet you. Almost everyone's dossier is available online. It's safe to presume that other people have looked you up on the Internet and already know about your accomplishments before they meet you. Thus, being humble scores even more points.

By the way As Facebook, LinkedIn, and other social media websites gain momentum, the people who use them offer more revealing, possibly inappropriate, and difficult-to-erase information to the public.[4] Make sure you protect your privacy by avoiding TMI (too much information) posts. For example, an employee who calls in sick but posts, "I'm off to the beach," might get busted by HR. Similarly, you don't want prospective employers or clients looking at your bachelorette party pictures online.

191. Ask open-ended questions. Questions are open-ended when they can't be answered with a simple yes or no. A few sample questions/comments to initiate conversation:

- Tell me more . . .
- That's interesting, how can you do that?
- I've always wanted to visit Hong Kong; what's it like?
- How did a chemical engineer end up in dental school?

192. Read up on current events and plan conversation starters ahead of time. Try regional topics, world news, sports, industry news, NOT politics, religion, or overly personal topics such as your messy divorce, your recent surgery, or your sister's last rehab visit. A few home-run conversation topics:

- Did you watch the Carolina/Duke game? (This is technically a yes/no question, but in March, you can name any two current U.S. college basketball rivals and spark great conversation. The key is to mention local teams and to find out if the other person follows sports before you ask.)
- Tell me about your family.
- I just read about the floods in Somalia.
- How did you get into this business?
- What do you think about (name a hot industry topic)?
- Who was your favorite speaker at the conference today?

193. Know when it's time to move on. If you feel a lag in the conversation or simply want to move on, politely excuse yourself and smile as you leave. You don't need to give an explanation; just say something like, "I enjoyed talking with you . . . please excuse me," or "I hope to see you again." You may want to employ the Wrap and Roll Technique we discussed in chapter 2.

194. Let other people talk. If you tend to talk a lot in social situations, plan moments to be quiet, such as when you have just answered a question or told a story. For example, if someone asks about your family, offer a quick anecdote, then ask the other person the same question: "What about you?" In reality, great conversationalists are simply good listeners.

195. Maintain a sense of propriety. If you're meeting to discuss business, make sure you meet in an office or other public location. Check with your organization for specific rules on entertaining customers and clients. Keep the conversation professional at all times.

MEETING THE MEDIA

A young financial consultant once told us a pertinent story about his first appearance on television. One morning a major news network contacted him and asked him to appear on an evening segment. He was so excited about being on national television that he spent the rest of the day emailing, calling, tweeting, and texting friends and family to invite them to watch. This financial whiz was very knowledgeable in his field and could have spent hours discussing this topic. Unfortunately, he wasn't given hours; he was given seconds. After he and the reporter exchanged niceties, he ended up with less than a minute to respond to a complex question. You can guess what happened next: he babbled out a disorganized, unrehearsed response that only partially answered the question, and then they went to commercial. This smart man learned a valuable lesson: prepare, prepare, prepare before you speak to the media. And have your answers in thirty- and sixty-second sound bites. Make only one call—to your mother. Trust us. She'll let everyone else know.

At some point in your career, you may have an opportunity to speak to the media, too. You may be asked to comment on a rare disease that you research or on the structural integrity of a building that collapsed. Many of the tips from this book will play into your ability to sound articulate for the media. We address a few additional nuances in the following tips.

196. Know your organization's policy on speaking to the media. Check with your company's legal and/or public relations department before you make any media comments.

197. Prepare ahead. If you're expecting a call for a quote or opinion, make sure you understand what the reporter wants and that you have additional information—such as journal articles and relevant data—handy. Take a moment to consider the kinds of questions the reporter will ask. Think about how to explain your topic in laymen's terms that would appeal to the general public. Make sure you can frame your knowledge into the context of the story.

> *By the way* If a reporter calls you for a comment, you don't have to say something immediately. It's okay to put the caller on hold while you collect your thoughts and notes or to ask if you can call back within a specific time frame.

198. Do your homework. If you have been invited to make a television, webcast, or radio appearance, research the program and the reporter who hosts it ahead of time.

199. Check out the show(s) on which you'll make an appearance. If time allows, watch or listen to the show where you'll be a guest. You'll get an idea of how long your interview will last, the kinds of questions the reporter may ask, and other hints about the experience. In chapter 3, we suggested that you read more. Similarly, if you anticipate future media appearances, watch and listen to appropriate shows on television, radio, and online. We highly recommend listening to shows on National Public Radio (NPR) as they often cover science and technology topics.

200. Ask the producer or contact person for tips. The show's staff wants you to do well. Don't be intimidated to ask them important preparation questions:

- What time would you like for me to arrive/call in?
- What is the goal of this interview?
- What do you think the reporter will ask?
- Do you have any insider advice for me?

201. Keep notes handy. As we mentioned, you can use notes with great success in a telephone conversation. If you're being interviewed over the phone or on the radio, make notes of what you want to say and jot down responses during the interview. If you're appearing on television, tuck these notes in your pocket.

202. Be aware of subtle differences in print versus broadcast media. When speaking to reporters who will quote you in a print or website story, carefully plan what you want to say. Grammar mistakes are more obvious in print. For example, most people will not notice the agreement error in this sentence when they hear it: "There's a lot of data on this problem." If they read it, however, they will likely see that the singular "There's" (there is) does not agree with the plural "data."

203. Choose your words carefully, especially when you're being interviewed on the radio. If listeners don't have a picture to watch, they will concentrate more on your word selection. Plus, radio stations such as NPR typically provide transcripts of their interviews online, where everyone can read and analyze what you said.

On a recent radio interview, an ob/gyn said, "Many women are not clued-in on the consequences of C-sections." This example is subtle but powerful, as the comment could backfire. Upon hearing it, we asked, "Is this doctor saying that these women are clueless?" Probably not. But the term "clued-in" implies that the information is readily available to all women, if only they paid attention. The doctor would have sounded more sensitive to his audience if he had said, "Many women are not aware of the consequences of C-sections." The word "aware" implies that their ignorance isn't their fault.

204. Be quiet after you answer a question. In broadcast media, reporters generally want sound bites. To ensure that your comments aren't overly edited, make your point in a clear, descriptive way that does not require excess explanation. For example:

Reporter:
"Dr. Zaret, how can our viewers change their diets to prevent heart disease?"
Dr. Zaret:
"I have three suggestions. First, eat at least five servings of fruits and vegetables each day; second, choose whole grains for breads, pastas, and rice; and third, try to exercise at least thirty minutes per day."

Notice how Dr. Zaret's answer is tightly crafted. It begins with an agenda and is parallel in structure, uses specific examples, offers numbering for transition, and makes the point in just a few words. It would be very easy for Dr. Zaret to offer unnecessary details that a producer would cut before airtime and unintentionally omit important points Dr. Zaret had wanted to make. (See chapter 3 for more information on parallel structure and other sentence structure concepts.)

205. After you make a media appearance, watch or listen to your performance and send a note of thanks to the people who made it happen. Watching or listening to your performance may be difficult at first. Once you get over your physical appearance and voice quality, you can pick up on things to improve next time. Thanking the publicist, producer, and reporter who helped you might get you an invitation to return.

WRITING AND DELIVERING PODCASTS

Podcasts, the online "radio shows" of today, and more recently VODcasts, their video-based equivalents, are becoming mainstream sources of communication. To date, teachers and professors have embraced Pod- and VODcasts most successfully. As their students enter the workforce, they'll bring the latest technology. For this reason, many people who never imagined they would record a lesson that could be heard and seen around the world will probably be creating Pod- and VODcasts within the next few years for multiple purposes.

University of Missouri technical business analyst Peter Meng writes, "Most likely Pod/VODcasting will not replace traditional broadcast radio or television, but become an intelligent extension of it, offering more variety to a significantly larger audience from an ever increasing number of content providers and producers, each with their own unique, highly-targeted revenue models."[5] Basically, this means that anyone can produce a "show." As with other forms of communication, however, the success of that show depends on the quality.

Experts in science, technology, and medicine have always embraced new technology as a way to teach others and enhance their research. If you don't already produce Pod/VODcasts, we recommend that you give them a try; they are a great way to showcase your work. As with any media opportunity, we advise that you seek approval from your organization's legal advisers before you begin. Here are a few production tips.

PREPARING FOR PODCASTS

Start with Podcasts. They require less equipment and technical expertise and offer an opportunity to "practice" your presentation skills with detailed notes at hand. One talented Podcaster, who also happens to offer great writing advice, is Grammar Girl. See the resources section of this book for her web address.

Freelance writer/editor Cathy Chatfield-Taylor summarizes the qualities of a good Podcast on her website, CCT Unlimited.[6] Here's a composite of her list with our commentary.

- Short (fifteen minutes or less). This is great advice since most of us have about a fifteen-minute attention span.[7]
- Conversational (features more than one person). You can also produce a great Podcast with only one speaker. Just make sure the speaker is lively and interesting.
- Moderated (identifies speakers and controls content). Make sure you have written permission to use other people in your Podcast.
- Instructive (gives take-away tips). Of course we loved this tip, as it's our credo!
- Glitch free (plays without interruption). This means you'll need to prepare and practice, practice, practice.
- Lively (with animated speakers). Please see chapter 4 for tips on developing strong delivery skills.
- Production-quality (using musical interludes and fade-outs).

Get the help and equipment you need. Once it hits the Internet, it's too late for a redo.

By the way We debated where to put this section on Podcasts: writing, meeting, or presenting? Although a Podcast is a recorded message where you "meet" the public, an important part lies in the writing. Be sure to refer to chapter 3 for tips on effective writing.

PREPARING FOR VODCASTS

To produce an Internet-worthy VODcast, you should invest in a professional videographer. Brilliant content and delivery can't overcome poor-quality video and sound. Check within your organization first. Then, depending on your budget, seek reputable videographers in your area. Our chapter on presentation skills covers everything you need to know on presenting in any setting. Grab tips from it and remember the following television- and VODcast-specific tips.

206. Wear simple clothing in solid colors. What you wear depends on your audience. If you're a professor recording a lecture for your students, dress as you would for class. If you're a professional explaining a product or idea, wear a suit. Choose flattering, subtle colors. Remember, computer video does not have the same quality as television and may distort patterns and colors. Avoid pinstripes, herringbone, busy prints, bulky clothing, and sparkling jewelry.

207. Choose simple words to describe your topic. In a VODcast, your audience will be focused more on the picture than on the audio. Make sure what you say is easy to follow and understand. (It will also be easier to deliver.)

6 Serve

To date, Amy had known nothing but success her whole life. At twenty-four, she had earned a PhD with honors in biochemistry. Following that, she was awarded a coveted two-year fellowship at a prestigious university. From there, she was hired as a medical liaison specialist (MSL) for a Fortune 500 pharmaceutical company, each year receiving "Exceeds Expectations" or "Outstanding" on her performance reviews. The company, recognizing Amy's potential, happily paid for her to attend an Executive MBA program. At the age of thirty, with a PhD and an MBA in hand, Amy was given a gem of a promotion: Director of New Business Development. Success! A lucrative salary. A prestigious title. A highly educated, productive team to lead. Nothing could stop her—or so she thought. But shortly after starting in her new leadership position, Amy began having trouble: deadlines were missed, projects went off track, and two of the most productive and long-term team members quit, saying they just couldn't work for someone like her. And for the first time in her life, Amy got a "Needs Improvement" at her performance review. She was flummoxed.

Of course, Amy thought her team members were the ones with the problem. "Why can't they meet these deadlines? What is it they don't get? Why aren't they adding any new ideas in our meetings?" Meanwhile, the team members were thinking, "She is totally unapproachable. She doesn't listen to a thing we say. What exactly is it she expects?"

This scenario represents a classic failure to communicate. Amy, like many of our seminar attendees, is exceedingly bright, well educated, and highly motivated, yet because of lack of training in many communication skills, such as active listening, collaboration, negotiation, and mentoring, she was unable to connect with her team and they with her. Unfortunately, Amy lasted less than eighteen months in the director position and found herself unable to make it as a team leader. She's currently an MSL for another company and taking courses in communication skills, hoping someday to get back into a leadership position.

These days communicating with employees, customers, colleagues, and others is more than "managing." Senior management officers in any organization expect employees to perform beyond what they are told to do. The kind of leadership required to communicate in this environment can best be described as serving. If you want to earn more money, step up to the next position in your career, and, in some cases, simply keep your job, you have to provide more value to your company. This means doing more than you're currently paid to do. After all, why should the organization pay you more if you haven't "earned" it?

If you think about it, most jobs create opportunities for you to serve other people. This doesn't mean that you become a personal assistant to every colleague or customer. Rather, it means that your job depends on their needing you. With this in mind, we have put some of our best tips in this last chapter and aptly titled it "Serve." When you see your work as a service to others, you immediately shift from having a job to making a career.

FACILITATE MEANINGFUL CONTRIBUTIONS

A software engineer who was managing a big project complained that his team members were not contributing equally. "It drives me crazy that not everyone takes this job seriously," he said. He was highly motivated and felt that some of the team members were ballasts instead of contributors. "Should I just ask them to step down?" he wondered.

The answer: no. If you expect everyone in the group to contribute the same passion and energy to the project as you do, you'll always be disappointed. Know that each person brings a different talent, life experience, or skill to the job. Each person will also bring different opinions, history, and baggage as well. As a project leader, manager, or supervisor, it's your job to find out what the team members expect to contribute, to take advantage of their expertise and contacts, and to deal with their preconceived notions to maximize their talents. If you master the ability to motivate all kinds of people on a variety of projects, you'll emerge as a leader in your organization. Take the tips we have given so far and assimilate them with this last chapter to become a leader in your field.

208. Be kind to others. It costs nothing and requires no skill. Your kind words, good deed, or thoughtful gift may even launch a cascade of positive gestures among others. A recent study by researchers from the University of California San Diego and Harvard University suggests that cooperative behavior spreads between people.[1] This ripple effect can have a wonderful positive impact on the corporate culture of your organization.

209. Respect diversity. We live in a global economy that seems to shrink a bit more with each new piece of technology. This ongoing metamorphosis brings us into contact with people from all cultures, religions, and backgrounds. For this reason, we must be mindful about how to communicate appropriately with all people. Use these opportunities to educate yourself about others. Avoid making comments on a person's race, religion, sex, national origin, age, disability, military membership or veteran status, sexual orientation, marital status, political affiliation, physical appearance, or mental health. (We're certain we left out some categories here, and we're counting on you to use good judgment on other potentially sensitive topics.) If appropriate, ask your colleague about cultural differences in communication and how you can adapt to his or her style.

210. Become a resource, particularly to a new employee. Thoughtfulness, graciousness, and kindness are never a waste of time. Building good relationships is a key to improving communication skills.

211. Start practicing the tips from this book today. Introduce yourself to someone you haven't met. Smile at everyone you see. Make small talk when you ride the elevator or pass in the hallway. People remember friendly, familiar faces. The gesture will come back to you again and again.

212. Become a mentor. This is one of the most rewarding things you can do in your career. You'll learn many things as you advise someone else: leadership, teaching, and supervisory skills. Plus, you'll develop a relationship that may pay big dividends in the future.

213. Cultivate a customer-focused environment. Many people have trouble seeing the benefit of helping others with their work. It's easy to think, "That's not my job." Instead, think of the other person as a customer. If you can help that person complete a task, it may expedite your goal. Soon you'll have the reputation of being a problem solver, the ultimate goal of any successful person.

214. Take a minute to "get a grip." It's okay to ask for a minute to think before you respond. Often, we're thrust into situations where people are making demands for our time or attention. Rather than snap at the person (and then have to apologize later), we recommend you ask for a moment. For example: "I have several things going on right now. Can you give me a minute to finish this so I can give you my full attention?"

215. Use good manners. Read up on appropriate table manners and social niceties. While it may seem stuffy or unimportant to know which fork to use and when, you'll be much more comfortable if you know what is correct when you dine with company executives or clients, especially people from other countries or cultures. Google "table manners" to find etiquette tips for dining and interacting with others.

SKILL BUILDER HOW TO SET YOURSELF UP FOR A
RAISE OR PROMOTION

We don't have a magic formula for getting to the next level in your career. We do know that to justify higher pay and additional responsibility, you must show that you have earned it. Here are a few basic tips that coincide with the overall premise of our book.

Do more than is required for your job, and document the extra work. The best way to justify more pay is to either generate income or save money for your organization. This is when you need to put on your critical thinking cap and showcase your ability to think conceptually and globally.

Point out how you have generated income or saved money for your organization. You can do this by documenting your work and sharing your success with your boss at an appropriate time.

Make your boss look good. Convincing your boss that you're ready to move up is the first step toward advancement. Don't try to skip rank and go over his or her head. Instead, ask what you need to do to move up.

Take care of "it" the first time. Nobody likes a procrastinator. When you establish yourself as the "go-to" person to get things done, you'll stand out with minimal effort.

Assume a leadership role. Take on an extra project or responsibility so you can learn the skills you need to obtain the new job while showing the boss that you can handle additional work.

Always have a positive attitude. People like to be around upbeat people. Smile and make people think that you're excited and enthusiastic about the work you're doing.

Use good judgment regarding ethics and integrity. If others lose your trust in these areas, you'll have a hard time earning it back.

216. Set yourself up as the consummate professional. How you dress, act, and work makes a huge impression on those around you. To set yourself up as a true professional (and the one who lands the project, job, or promotion), remember the following.

▪ **Dress appropriately.** You may not think that dress has anything to do with communication. In fact, your appearance is the first thing you convey when you meet someone. Make sure that what you wear is appropriate for your work environment and corporate culture. If in doubt, check with your boss or a human resources person for a dress code for your organization. Always err on the side of conservative.

▪ **Act appropriately.** When at work, you should work. Don't text your friends, gossip around the water cooler, groom yourself at your desk, use foul or unprofessional language, check personal email or social network sites, or shop online. If nobody mentions it, you may think nobody notices. On the contrary, people do notice and it could prevent you from landing that promotion or raise you desire.

■ **Work appropriately.** Make sure you always meet deadlines and obligations, without excuses or complaining. Show others that you're reliable and competent through your work ethic.

By the way People do business with people they like and trust. That is why your neighbor, rather than a telemarketer, will call to ask you for help with a fundraising campaign. Establishing trust is much easier than it sounds. The most important thing you can do is always behave in a professional manner.

217. Apologize. If you botch a project or say something stupid, apologize. Most people are willing to give you the benefit of the doubt if you show accountability. A good apology doesn't include a litany of excuses. It's brief and sincere. Of course, knowing that you can just apologize later does not give you permission to take advantage of others.

218. Avoid correcting your colleagues in public. A hospital director was taking an important group on a tour of the facility. While rattling off statistics about the organization, he misspoke. A colleague interrupted his speech to clarify the numbers. You can probably guess what happened next: the director was embarrassed and fumbled through the rest of the tour red-faced. We must ask: Did anyone gain anything by knowing about the mistaken number? Couldn't the colleague have mentioned it during a break so the director could correct himself later? Even in the name of "getting the information right," this colleague appeared to have only one motive: to embarrass the director. Not the best move to anyone who witnessed it.

SKILL BUILDER HOW TO DEAL WITH AN UPSET PERSON

A client called his financial adviser in a huff because the client read in the *Wall Street Journal* that a stock they agreed to purchase tanked. "I can't believe you suggested I buy this!" the client yelled. The financial adviser took a deep breath and prepared to respond.

Think about the last time you were upset. What did you want from the person responsible? You probably just wanted someone to listen and show empathy. Most upset people don't expect miracles. They simply want to be heard. With good communication skills, you should be able to calm the person down and negotiate a solution. Here's how to give the person what he or she wants without compromising a solution or wasting time.

Remain silent and calm. This will give the angry person an opportunity to speak uninterrupted. Most upset people want to vent. The financial adviser should let the client express all of his concerns before trying to placate him.

Echo what has been said. Get the facts, timeline, and people involved. Once the other person has stated the case, repeat back what he or she said in summary. (This takes great concentration, so be prepared to really listen, not plan what you'll say in response.) For example, "Let me make sure I understand . . . you would like to see . . ." In one of our seminars, an account manager from a software technology company called this step "cushion and clarify." She explained the "cushion"

part as showing empathy and the "clarify" part as an opportunity to find out more and pinpoint the problem.

Make an intention statement. After listening to the speaker
and repeating what was said, resist the desire to launch into
an excuse, solution, or debate. Instead, make sure you show
empathy and ownership by saying, "We didn't mean for this
to happen. I'm so sorry." Apologize even if you're not responsible for the problem. To the upset person, you represent the
organization and you should apologize. This step will carry
you toward a solution without angering the other person
further. After all, most upset people simply want an acknowledgment and an apology.

Ask for a proposed solution. "What would you like to see
happen?" You'd be surprised by the simplicity of the answer.
Most people will be satisfied with the apology. Those who ask
for more will probably want something that you can easily
accommodate. The financial adviser could use this step as an
opportunity to explain the details of the purchase and create a
plan to recover.

If the situation becomes out of hand, you may need to take a
different approach. One of our clients, a successful physician,
had an irate patient storming around the reception area, caught
in a tirade over a bill she had received from his clinic. The
patient took full advantage of the packed house and performed
with great gusto, yelling, cursing, and otherwise acting irrational. The receptionist was appalled to the point of speechless-
ness. The waiting patients were shocked yet curious to see how
things would unfold. Out came the physician. He calmly

stepped into the reception area, asked the woman to join him in a private office to discuss the problem, and motioned to a hallway. The woman, distressed that she was going to lose her audience, continued with the performance. The doctor smiled warmly and calmly said, "I'm not going to let you talk ugly to me." But the woman persisted with the spectacle. The doctor said again, "I'm not going to let you talk ugly to me." Yet she continued. He kept repeating his mantra, leaning in gently and almost whispering, "I'm not going to let you talk ugly to me." Finally, realizing the futility of her efforts, she retired to his office to discuss her bill in private.

Occasionally we all feel the way this woman felt: angry, helpless, and anxious to have a voice. Later she probably felt ashamed of her behavior. No one wants to lose control of the situation. This brings us to several great tips for dealing with irrational people.

219. Show empathy. Try to understand the other person's perspective. If you knew that the upset patient's husband has been sick for months with terminal cancer, that she just lost her job, or that her mortgage payment is overdue, would you be more empathic? Assume there is a back story and make every attempt to validate the upset person's feelings and give him or her the empathy and respect he or she craves.

220. Use humor to lighten the mood. Sometimes a little laughter can help diffuse a tense situation. For example, "I'm so sorry about this misunderstanding. I was coaching my six-year-old daughter's softball team this morning, and I may have suffered a touch of heatstroke or maybe it's just plain exhaustion. Sorry!" The key to making fun in a difficult situation is making sure you're the butt of the joke, not the other person. Be careful with humor as off-color jokes can make things worse.

221. Take a break. Some conversations simply need a break. If you find yourself caught in a heated discussion or dealing with a difficult person, try changing the subject or postponing the conversation. The best way to change the subject is to use the Wrap and Roll Technique we mentioned in chapter 2. First you *wrap* the conversation by echoing the other person's complaint: "I understand why you're upset about this health care benefit change." Then you *roll* to the next topic: "I'm also worried about the new policy. I remember when they switched pharmacies five years ago. It was difficult at first, but now I love it." Apply this same technique to postpone the conversation. *Wrap:* "I really want to hear all of your concerns." *Roll:* "Unfortunately, I have a meeting/phone call/appointment. Can we discuss this later? (Listen for answer.) When is convenient for you?"

222. Remove the audience. Some people enjoy the attention they get from being upset in front of others. An easy way to fix this situation: ask the upset person to step to a private location such as an office or conference room. When the audience is gone, the show is over.

223. Offer a bit of personal information. One way to win over anyone is to show you truly understand. The easiest way to do this is with self-disclosure: "I understand; we have a history of breast cancer in my family." By doing this, you not only show empathy, you instantly deepen the relationship by revealing something presumed to be private. The key to self-disclosure is making sure the information is appropriate. For example, ask yourself, "Will revealing this information make the other person uncomfortable?" Always protect the personal information that others share with you by not passing it on to others.

224. Resist the desire to reply in kind to inappropriate behavior. It's natural for humans to respond to how we're treated with the same behavior, particularly when someone treats us poorly.[2] Think of psychiatrist Walter Cannon's Fight or Flight Response theory, which leads to the "I was backed into a corner" justification of today. Avoid spewing back at people who are unkind, unprofessional, or just plain mean. Remember the reverse of this is also true: people tend to react to you in a way that corresponds to your behavior. These are occasions when you must ask yourself, "What is there to gain by 'winning' this argument?"

By the way You can use body language to assist you in communicating with upset people. A few ways are

- **Eye contact and nodding** show empathy and acknowledgment.
- **A "stop" hand.** Hold your hand up, with the palm out, to say, "stop."
- **The "just a minute" finger.** Close your hand and point your index finger upward (palm side out) to say, "Please wait just a minute." This is very effective if you're otherwise occupied with a phone call or speaking with someone else. Be sure to smile when you give this signal so you don't appear rude.
- **Facial expressions.** Be aware of what your expressions say for you. A smile can show empathy but can also appear condescending if overdone. Furrowed brows look serious. A slight frown can express unhappiness about the situation.

SKILL BUILDER HOW TO DEAL WITH DIFFICULT PEOPLE AND DIFFICULT SITUATIONS

We have all been around people who are difficult: the colleague who rants about her dissatisfaction with work, hoping you'll agree and propel the conversation by adding your grievance; the boss who shares too much information about other employees when they are absent from the room; the peer who continually tells inappropriate jokes and pushes the line with unprofessional office behavior. These situations are particularly difficult when the offending person is a coworker or colleague. The following tips answer some of the common questions we hear from our seminar participants. Often the best advice came from other members of the audience!

How do I respond to the co-worker who complains about management incessantly?

First, reserve comment, and then get out of the conversation as quickly as possible. Even if you don't say anything, you don't need to be associated with the griping. Someone might infer that your presence equals agreement.

Remember the Wrap and Roll Technique: summarize what the co-worker said, show empathy, and move on to other topics. It's always appropriate to remove yourself from the discussion by

- Referring others to HR for solutions: "I think you may want to speak to Lin in HR about this . . ."
- Excusing yourself from the conversation: "I'm sorry, I have

a pressing deadline," or "I'm not comfortable talking about this."

By the way If you're caught in an uncomfortable situation where the other party is giving too much information, saying things that are inappropriate or unprofessional, or otherwise wasting your time, it's fine to say, "I appreciate that you feel comfortable sharing this information with me. I'm just not comfortable being your confidant."

What do I do about the boss who writes poorly and asks me to "edit" his work all the time?

Remember that you work for your boss, and when you make him look good, you look good, too. If you have another pressing deadline, ask him to give you a deadline for his project and guidance on how to prioritize assignments. Use "I" statements to introduce a question or potential problem: "I notice that you . . ." Another good sentence to use is "Here in the document where you say _____ you might want to say _____ instead."

How do I "supervise" people who don't report to me?

If you're placed in a situation where your co-workers are helping you with a project, or working with you regularly, you should discuss your options and responsibilities with your supervisor, and plan how to accomplish the project with the help of the group. Remember, this is a great opportunity for you to show leadership and collaboration skills as a project manager.

To build support and lead others to complete a project,

- **Create buy-in.** Discuss the project with the group and ask for help to solve problems and meet deadlines.
- **Answer why.** Make sure you offer the group the reasoning behind the project, for example, "ABC hired us to develop a new cable system for their plant. I invited each of you to be on this project team because you have a specific technical expertise."
- **Ask for help.** Most people are willing to help if asked. Don't assume that others can't or won't help with your project. A great way to start a new team meeting is to introduce everyone individually and share what talent each person brings to the group. For example, "To my left is Brenda Wang. Brenda has been a network analyst for over ten years. I asked her to help with this project because she is a PowerPoint™ maven."
- **Explain the benefit.** Make sure you frame the project in terms that show a benefit to all involved parties, for example, "I think this project will help us further our goal of..."

How do I bring up a difficult subject?

We always recommend using "I" statements to introduce touchy subjects. We also recommend that you avoid naming others, placing blame, or pointing out personality flaws.

For example:

Instead of
"David told me..." or "I hear..."

Say

"I understand that . . ." or "I noticed . . ."

What do I say to the co-worker who thinks it's my job to assist and inform her of every change at work?

An audience member told us that she offered to help a new employee get adjusted to the job. The new employee was gracious and grateful. After a few months, the audience member started to wean the new employee so that she could establish herself. Meanwhile, the new employee ran into a small problem, returned to the mentor in a huff, and said, "Why didn't you tell me about this?" The mentor was surprised by the outburst and calmly apologized. The relationship was saved and the new employee was officially weaned.

Sometimes colleagues will perceive that you should report every company detail to them, even when the detail is privileged information or irrelevant to them. There is no advantage in telling these people why the information is unavailable. The best approach is simply to say, "I'm sorry I wasn't allowed to share that," and move on to another subject.

How do I deliver bad news?

No one wants to have this job, yet we all have to deliver bad news occasionally. A few things to consider:

- **Timing.** Make sure the timing fits the information and audience. Certain messages can't wait for "the perfect time," while others must. Be sure to allow time for the recipient to react or comment without interruption.

- **Location.** If you think the news will generate an outburst or embarrassing moment, make sure you're in a private location.
- **Word choice.** If there was ever a need to practice a speech, this is it! Think carefully about how you should frame the message. Consider the audience, what is important to them, and how they might react.
- **Delivery.** This is when you need to pull out all of the great delivery techniques from chapter 4 to show empathy: eye contact, body language (leaning in, appropriate touching), and facial expressions. Be sure to use calm, soothing speech by speaking at a slightly slower rate and lowering your voice.

By the way Multiple studies show that humans respond positively to appropriate touch with greater cooperation and compliance.[3] Appropriate ways you can touch others when communicating bad news are a shoulder pat, handshake, or light forearm touch.

SKILL BUILDER HOW TO MANAGE AND EXPEDITE PROJECTS

Many project managers complain that keeping their projects moving forward can seem even more challenging and time-consuming than the work itself! Instead of implementing the new ideas generated by their project, these project managers fill their time organizing schedules, corresponding with co-workers, and checking to make sure other people are doing their share of the work. Yet project management is an integral part of most jobs. So how can you do it with grace and ease? Check out these tips.

Build relationships from the start. We have already said that people do business with people they like and trust. Start your next project with a short teambuilding exercise. In our seminars we often ask all participants to introduce themselves, tell us about their jobs, and explain why they came to our seminar. You can also devote a larger amount of time to storytelling exercises including details about your first job, your favorite teacher, something obscure about you, and so on.

Set up a mutually agreeable plan with specific goals and attainable deadlines. This will help everyone take ownership of the project.

Set regular meetings either in person or via conference call. When the group agrees on a weekly conference call or monthly meeting, everyone shares the work deadline it creates and the responsibility to attend.

When meeting, always offer the group an agenda ahead of time and a summary of assignments afterward. As the project manager, you can delegate a portion of this work. (See chapter 5 for more tips on effective meetings.)

Create an environment of accountability. Remember that as the project manager, you're ultimately responsible for the completion of the project. Your team will count on you to help them set and clarify priorities, collaborate for solutions to problems, and remain steadfast in your decisions. They will expect you to keep the project moving and to handle the difficult people and situations that occur.

▌ **Anticipate problems and plan potential solutions.** We all know that problems will arise; they key is being prepared to ▌ handle them.

▌ **Speak to individuals for the "rest of the story."** If you have ▌ trouble keeping the team on task, speak to individuals privately to find out why they are not productive and use performance ▌ mance management discussions to hold them accountable.

OTHER WAYS TO SERVE

225. Volunteer. One of the easiest ways to enhance your skill set is to volunteer your time. Contact your local Chamber of Commerce, school system, or not-for-profit organization, and offer to help with a project. You'll make great business contacts while testing a new skill in a nonthreatening environment.

226. Focus on the positive. In these times of constant change with buyouts, mergers, and foreclosures, it's easy to dwell on the negative. Try to always recognize the positive in your staff and the work environment because others will turn to you as a model of behavior and attitude.

227. Take ownership of your work. Be proud of your work and prepared to make it better. If your boss or colleague comments that your work should be different from what you think, ask why. Avoid explaining your methods or opinions ad nauseam. Instead, take note of what you need to fix and work toward a better end product. Ultimately, you'll impress your peers and your superiors if you're willing to admit mistakes, fix them, and move forward.

228. Delegate. As you climb the ranks in your organization, you'll quickly realize that it's impossible to micromanage every project for which you're responsible. You must delegate tasks to the people who work for you. Make sure that they are informed and trained on what needs to be done, that you allow them to do the work, and that you monitor their progress periodically.

229. Be organized. Throughout this book, we have suggested making lists, planning ahead, and practicing before you communicate. Great communicators take advantage of strong organizational skills: preparing comments before a meeting, taking a moment at the end of the day to assess what needs to be completed tomorrow, evaluating how a project is progressing, and changing course as needed. Solid organizational skills will serve you well as you become a leader in science, technology, or medicine.

230. Give yourself a break. You can't expect to finish reading this book and automatically have perfect communication skills. Just like any great skill, these require practice. Try implementing one tip per day. Keep this book handy and pull it off the shelf the next time you have to counsel an employee, write a letter, or give a presentation.

SKILL BUILDER HOW TO RESPOND POSITIVELY TO COMPLAINTS AND SUGGESTIONS

Complaints and suggestions can actually be great tools for your organization. The complainer may offer you insight on a problem that you didn't know existed, such as an inventory shortfall. If you catch a problem, alert your boss, and then fix it, you'll reap the benefits. The key to dealing with people who come to you with a problem isn't to take it personally. Typically complainers are not upset. They usually want to find a solution, such as direction on how to open a file from your website. Meanwhile, your goal is to get back to the tasks that are important to you. Here's how to meet both objectives.

Thank the person and explain why you're glad to hear the information. Even if ultimate responsibility for a problem lies in "the system" or another department, if someone has chosen to tell you about it, then, in that person's opinion, you're responsible for it.

Apologize for the problem and help the complainer solve it. This is your opportunity to explain who is ultimately responsible and how you'll contact that person. Start with an apology (even if the problem isn't your fault). For example: "I'm sorry you're having a problem." This shows empathy and good listening skills.

If appropriate, direct the complainer to the person who can help: "I think you may need to speak to our webmaster, Nuala Kennedy. May I give you her contact information?"

Collect all the information you need to fix the problem and, if necessary, send it to the responsible party. Although you may be able to deal with the problem immediately, you still need to pass the information to the person who can fix it permanently.

Correct the mistake as quickly as possible. If you passed the problem to someone else, make sure that person takes action.

Follow up to make sure everyone is satisfied with your actions. If you're not the person who can solve the problem, it might seem unnecessary to follow up. On the contrary, that's exactly *why* you should follow up. Both the complainer (who may be a customer, colleague, or boss) and your peers will see you as a problem solver. As we said earlier, this is the ultimate tag you want associated with your name.

SUPERVISING OTHERS

Besides being responsible for our own work, many of us are also charged with overseeing the work of others. This task can be one of the most challenging parts of a job, especially if you're embarking on it for the first time. Many of the tips in the previous chapters of this book apply to supervising others. In addition, note the advice in this section.

231. Offer clear directives to subordinates. You can almost never go wrong with specific expectations for those who report to you. You don't have to be "bossy" to make this happen; just be specific. For example, a supervising pharmacist might explain telephone protocol to a pharmacy technician in the following terms: "The phone rings all the time here in the pharmacy. To make sure we answer every call in a timely fashion, the pharmacy techs man the phone. Your job is to catch the phone, help the customer by answering as many questions as you can, then pass it on to me or another pharmacist if necessary."

232. Don't micromanage your employees. Offer them appropriate training and decision-making skills and let them do their jobs. Most people will tell you that when they feel micromanaged, they become disengaged and less productive on the job.[4] Advise them and give them permission to handle problems on their own. This is a great time-saver and morale-booster. If appropriate, seek their advice on issues that pertain to them, for example, "Bill, what do you think about the new policy?" or "How do you think we can fix this problem?" Employee ownership of a project will propel it to completion.

233. Invite suggestions from others. Don't get caught up in hierarchies or titles. Talent is all around you. Try to always surround yourself with smart people. Let people know that you're open to their ideas and value their input. Many great ideas come from what seem to be the most unlikely places within your organization.

234. Don't be afraid to hire people smarter than you. Successful leaders are never threatened by people who know more than they do. In fact, they prefer to hire people who have skills they don't. True leaders are always looking for other leaders.

235. Don't wait for a problem to arise. If you're new to a supervising job, meet with your employees to outline your expectations. If you have hired a new employee, make sure you train him or her immediately. You'll save valuable time later and prevent possible mishaps and miscommunications.

By the way One of the easiest ways to deal with difficult situations is to prevent them. Here are three easy ways to do this.

▪ **Keep track of recurring problems or complaints.** If you notice a recurring problem, track it on a notepad or spreadsheet, discuss it with colleagues, and examine the data so you can figure out and fix what is causing it.

▪ **Empower the people who work for you.** For your employees to be able to do their jobs well, they need to be allowed to prevent problems, deal with difficult people, and manage tough situations when they occur. Offer training and guidance on how to deal with common scenarios and potential problems. For example, create a protocol for dealing with upset customers, make a checklist for completing a task, and give your employees "permission" to solve problems on their own.

▪ **Create a customer satisfaction survey.** Your employees will automatically improve service if they know they are being evaluated. Plus, your customers will appreciate your efforts when you ask for their opinions.

236. Follow your organization's policies to the letter. Make sure that you and your employees understand and follow the organization's policies and procedures. You can prevent lots of misinterpretation by covering this early in your relationship.

SKILL BUILDER HOW TO PREPARE FOR A DIFFICULT
PERFORMANCE REVIEW

Anyone who has ever managed people will tell you the hardest part is giving criticism. Here's how to do it gracefully, effectively, and efficiently.

Choose the right time and place. Generally speaking, the best time to deliver bad news is at the end of a shift, at the end of a week. This choice allows a graceful exit for you and the employee if needed. (Based on your organization's human resources policy and corporate culture, you may choose a different time.) The best place to deliver both praise and criticism is in your office, behind closed doors. This setting preserves the other person's privacy and often diminishes the "call to the office" in the eyes of co-workers.

Follow company protocol and document everything in writing. If you're firing an employee, make sure you use the organization's protocol to the letter. If you fear the other person will make false claims against you later, ask an HR person or your boss to join you. Make sure you document everything in writing and ask the employee to recognize your documentation with a signature.

Start by asking the person if he or she is open to some constructive criticism. In this way, you'll have agreed to more receptive communication. In addition, ask the person to evaluate his or her own work. One way to transition into your main objective is to say, "There's something we need to discuss."

Make sure you have specific, observable behaviors and examples to back up your claims. Avoid criticizing personality traits. Instead, use this time to point out habits that conflict with the person's job description and productivity.

For example:

Instead of
"You're so quiet in meetings. I need for you to speak up and give more of the supporting data for the project."
Say
"I notice that you didn't have much to say about the project in the meeting." (Pause and wait for a response.) "I'd like to see you speak up and share more of your research on wind patterns as it relates to the project."

> *By the way* The key to delivering feedback, advice, or general observations is to make sure you sound sincere and not sarcastic. People associate sarcasm with an emphasis on certain words in a sentence. In the previous example, emphasizing the word "say" in the first sentence might sound sarcastic.

Recognize efforts made to prevent this outcome. Most people are conscientious enough to want to do well. Ask the employee, "What steps did you take to prevent this situation?" and "What would you do differently next time?" Instead of reprimanding, you're coaching the employee to handle difficult situations more effectively.

Ease the burden of blame with "I" statements. "I" statements allow you to shift the focus onto the problem rather than the person.

For example:

Instead of
"You've been late three days in a row. What's going on?"
Say

"I notice you've been late three days in a row. Is everything okay?"

Show empathy without appearing insincere. This is where you need to watch for the tone in your voice. You don't have to coddle others to show empathy. Say something like, "I can see that you're swamped. What needs to be done next?"

Give the other person a chance to speak. If you really want answers, be quiet. After you ask another person a question, resist the temptation to fill the void with guesses on the answer. Instead, wait for the other person to speak. Eventually, he or she will break the silence.

Help generate ideas for solutions. Say, "What can we do to fix this?" or "How can the team help?" Good managers seek solutions. Great managers help others find them.

Give others the benefit of the doubt. Someday you'll need a return of this favor.

237. Put forth extra effort even when there is no obvious gain. Many times the benefit of your actions may not be clear or forthcoming. For example, the co-worker you help may not be in a position to repay for another six months. Eventually the reward for your efforts will be evident.

238. Ask, "What may I do to help?" This simple line can open many doors for you. When you ask this question, you show that you have an interest in helping, which makes you appear indispensable. If you're able to complete the task needed, you actually become indispensable.

239. Give people clear instructions and ownership of their projects. Always publicly recognize those who have done a good job. Occasionally send handwritten notes thanking people for their contribution and hard work.

240. Let others shine in your presence. Never take credit for the work your team or one of its members has done. People often get to the top because the people around them or below them helped them get there. Plus, a loyal team will protect and defend you when necessary.

241. Don't take advantage of your position. Remember that other people are watching your choices. If you're supposed to take a one-hour lunch, don't stretch it out to two hours because you're the boss. Your subordinates will look to you for an example (or an excuse) of what is appropriate.

242. Allow others to win. Be someone's champion. You'll have a lifelong advocate. For many people, compromising sounds like giving up the win. In fact, when you strategically give something up, you make room for something better.

For example:

Let's say a colleague rushes into the copy room where you're using the equipment and asks, "When will you be done?" You still have quite a few copies left, and, since you were there first, it only seems fair for the other person to wait. But what if your copies can wait? You may want to give up the copier to the other person. You might say, "I know how you feel . . . I sometimes get caught in deadlines, too. Why don't you go ahead and let me know when you're done?"

243. Don't push others down on your way to the top. If you have to put others down to get to the top, you won't be prepared once you arrive. If your promotion is in the same organization, the people you pushed down to get there will remember. Ultimately you'll have tarnished your reputation and hindered your chances to make the next step up.

244. Lead by example. For most upper-level jobs, your success depends on your subordinates. Your boss expects you to guide your employees to do well, leading you to benefit from their successes. Unfortunately, you'll also experience the disappointment of their failures. Supervising others can be a challenge to even the most experienced managers. Although you can't control your employees' behavior, you can control your own. Start by communicating effectively with the people who work with you to show them the example of leadership. After all, imitation is the form of modeling you used when you first learned to communicate as an infant.[5]

245. Become a lifetime learner. In order to master the concepts in this book, you must become a lifetime learner. People who are lifetime learners welcome new assignments and challenges in their careers. They see the presentation at the board meeting as an opportunity to show the vice president their knowledge base. They know that the business plan they have just been asked to write will be a great addition to their résumé. They realize that completing the tasks that are "not part of their job description" will be the keys to their promotion.

COMMUNICATE WITH US

Thank you for reading this book. As we have mentioned repeatedly, no matter what mode of communication you use, your style, word choice, and message should reflect what you truly want to communicate.

As lifetime learners, we are always looking for new tips and examples. We'd love to hear from you. Please send your stories and questions to Stephanie@listenwritepresent.com or Deborah@ listenwrite present.com, and visit our website, www.listenwrite present.com, for more tips.

Notes

Preface

1. Mintz C. Transcending the transition from academia to industry. *Science.* May 14, 2004. Available at http://sciencecareers.sciencemag .org/career_magazine/previous_issues/articles/2004_05_14/noD OI.7805445794410933401. Accessed Dec. 28, 2010.
2. Tay L. IT professionals take on consulting skills: masterclass aims to tackle a lack of soft skills. *Computerworld.* Jan. 29, 2007. Available at http://www.computerworld.com.au/article/173508/it_professionals _take_consulting_skills. Accessed Sept. 4, 2009.
3. Bartoo C. New curriculum shines at school of medicine. *Reporter.* Aug. 14, 2009. Available at http://www.mc.vanderbilt.edu/reporter/ index.html?ID=7455. Accessed Sept. 4, 2009.
4. Staff-Parsons M, Pullman W. Transforming clinicians into industry leaders. Pharmexec.com. Available at http://license.icopyright.net/ userviewFreeUse.act?fuid=NDc2OTIiMw%3D%3D. Accessed Sept. 4, 2009.

Chapter 1. Plan

1. Stein K. Shop faster. *New York Times.* April 16, 2009. Available at http: //www.nytimes.com/2009/04/16/opinion/16stein.html. Accessed April 16, 2010.

2. Ophir E, Nass C, Wagner A. Cognitive control in media multi-taskers. *Proceedings of the National Academy of Sciences*. Sept. 15, 2009;106(37):15583–15587.

3. Huitt, W. Maslow's hierarchy of needs. *Educational Psychology Interactive*. Valdosta, GA: Valdosta State University; 2007. Available at http://www.edpsycinteractive.org/topics/regsys/maslow.html. Accessed April 14, 2010.

Chapter 2. Listen

1. MacDonald K. Patient-clinician eye contact: social neuroscience and art of clinical engagement. *Postgraduate Medicine*. 2009;121(4):136–144.

2. Diamond S. *Getting More: How to Negotiate to Achieve Your Goals in the Real World*. New York: Crown; 2010:69.

Chapter 3. Write

1. Manivannan G. Technical writing & communication: what & why? July 19, 2009. Available at http://www.usingenglish.com/teachers/articles/technical-writing-communication-what-why. Accessed Aug. 26, 2009.

2. Eleven reasons why manuscripts are rejected. San Francisco Edit. Available at http://www.sfedit.net. Accessed June 29, 2010.

3. Bordage G. Reasons reviewers reject and accept manuscripts: the strengths and weaknesses in medical education reports. *Academic Medicine*. Sept. 2001;76(9):889–896.

4. Roland M. Publish and perish: hedging and fraud in scientific discourse. *European Molecular Biology Organization Reports*. May 2007; 8(5):424–428.

5. Barnard S, and the Health Care Communication Group. *Writing, Speaking, & Communication Skills for Health Professionals*. New Haven: Yale University Press; 2001:53.

6. Guidelines for creating plain language materials. Available at http://www.centerforplainlanguage.org/aboutpl/guidelines.html. Accessed May 24, 2010.

7. Gopen G, Swan J. The science of scientific writing. *American Scientist*. Available at http://www.americanscientist.org/issues/feature/the-science-of-scientific-writing/5. Accessed Jan. 29, 2009.
8. Zinsser W. *On Writing Well.* 7th ed. New York: Collins; 2006:71.
9. Fowler R, Hoffman N, Kolata D, Shah A. Building Learning with Technology website. University of Maryland. 2003. Available at http://www.education.umd.edu/blt/unit/pgcc/PersuasiveWords.htm. Accessed April 21, 2010.
10. Rowse D. Writing blog content—make it scannable. Available at http://www.problogger.net/archives/2005/08/19/writing-blog-content-make-it-scannable/. Accessed April 19, 2010.
11. Sharma M. Top 10 overused buzzwords in LinkedIn profiles. LinkedIn blog. Available at http://blog.linkedin.com/2010/12/14/2010-top10-profile-buzzwords/. Accessed Jan. 9, 2011.
12. Pease B, Pease A. *The Definitive Book of Body Language.* New York: Bantam; 2006:9.

Chapter 4. Present

1. Farnham S. Social psychology online. American Psychological Society. Dec. 2000. Available at http://www.psychologicalscience.org/observer/private_sector/farnham.html. Accessed Sept. 23, 2009.
2. Beware of dissatisfied customers: they like to blab. Knowledge@wharton website. March 8, 2006. Available at http://knowledge.wharton.upenn.edu/article.cfm?articleid=1422. Accessed Feb. 2, 2011.
3. Blum D. Face it! *Psychology Today.* Sept. 1, 1998. Available at http://www.psychologytoday.com/node/25371. Accessed Sept. 18, 2009.
4. Farnham, Social psychology online.
5. Lambert D. *Body Language 101.* New York: Skyhorse Publishing; 2008:24.
6. Jones D, Motluk A. Eight ways to get exactly what you want. *New Science.* May 7, 2008: 2655. Available at http://www.newscientist.com/article/mg19826551.400-eight-ways-to-get-exactly-what-you-want.html. Accessed April 16, 2010.
7. Mehrabian A, Wiener M. Decoding of inconsistent communications. *Journal of Personality and Social Psychology.* 1967;1:109–114.

8. Blum, Face it!
9. MacDonald K. Patient-clinician eye contact: social neuroscience and art of clinical engagement. *Postgraduate Medicine.* 2009;121(4):136–144.
10. Blum, Face it!
11. Lieb S. Principles of adult learning. Honolulu Community College faculty development web page. Available at http://honolulu.hawaii.edu/intranet/committees/FacDevCom/guidebk/teachtip/adults-2.htm. Accessed April 22, 2010.
12. St James D. *Writing and Speaking for Excellence: A Guide for Physicians.* Boston: Jones & Bartlett; 1996:240.
13. McWade J. Design talk 19: five design ideas. *Before & After.* Available at http://www.bamagazine.com/. Accessed Sept. 8, 2010.
14. O'Connor SL. Creating effective slides. *American Medical Writers Association Journal.* 2010;25(2):59.
15. Deeb S, Motulsky A. Red-green color vision defects. *GeneReviews.* Sept. 19, 2005. Available at http://www.ncbi.nlm.nih.gov/bookshelf/br.fcgi?book=gene&part=rgcb. Accessed May 28, 2010.
16. Learning Resources Center. Effective Reading Techniques. University of Utah School of Medicine. Available at http://medicine.utah.edu/learningresources/tools/reading.htm. Accessed May 28, 2010.
17. Kurtus R. Samuel Clemens (Mark Twain): age 60 to death at 74. School for Champions website. Available at http://www.school-for-champions.com/biographies/marktwain4.htm. Accessed Sept. 13, 2009.
18. Rossiter M. Narrative stories in adult teaching and learning. *Education Resources Information Center Digest.* 2003;4. Available at http://www.ericdigests.org/2003-4/adult-teaching.html. Accessed April 23, 2010.

Chapter 5. Meet

1. Bielaszka-DuVernay C. Take a strategic approach to persuasion. *Harvard Business Review OnPoint.* Spring 2011;20.
2. Waggener Edstrom Worldwide. Survey finds workers average only three productive days per week. *Microsoft Press Pass.* March 15, 2005.

Available at http://www.microsoft.com/presspass/press/2005/mar05/ 03–15threeproductivedayspr.mspx. Accessed Sept. 22, 2009.

3. Nass C, Raeburn P. Multitasking may not mean higher productivity. [transcript]. National Public Radio Science Friday. Aug. 28, 2009.

4. Rosen J. The web means the end of forgetting. NYTimes.com. July 21, 2010. Available at http://www.nytimes.com/2010/07/25/maga zine/25privacy-t2.html. Accessed Aug. 2, 2010.

5. Meng P. Podcasting and VODcasting: a white paper. University of Missouri. March 2005. Available at http://www.wssa.net/WSSA/Soc ietyInfo/ProfessionalDev/Podcasting/Missouri_Podcasting_White _Paper.pdf. Accessed April 22, 2010.

6. Chatfield-Taylor C. What makes a good podcast? CCT Unlimited website. Available at http://cctblog.typepad.com/cctnewsblog/2006/ 04/what_makes_a_go.html. Accessed April 22, 2010.

7. Middendorf J, Kalish A. The "change-up" in lectures. Campus Instructional Consulting. Indiana University TRC Newsletter, 8:1. Fall 1996. Available at http://www.indiana.edu/~teaching/allabout/ pubs/changeups.shtml. Accessed Dec. 28, 2010.

Chapter 6. Serve

1. Fowler J, Christakis N. Cooperative behavior cascades in human social networks. *Proceedings of the National Academy of Sciences.* Jan. 25, 2010. Available at http://www.pnas.org/content/early/2010/02/ 25/0913149107. Accessed Aug. 29, 2010.

2. Keysar B, Converse B, Wang J, Epley N. Reciprocity isn't give and take: asymmetric reciprocity to positive and negative acts. *Psychological Science.* Dec. 2008;19(12):1280–1286.

3. Adler R, Rosenfeld L, Proctor II R. *Interplay: The Process of Interpersonal Communication.* New York: Oxford University Press; 2010: 190–191.

4. Fracaro K. The consequences of micromanaging. *Contract Management.* July 2007:4–8.

5. Meltzoff A. Born to learn: what infants learn from watching us. Available at http://ilabs.washington.edu/meltzoff/pdf/99Meltzoff_ BornToLearn.pdf. Accessed May 27, 2010.

Recommended Resources

Books on Interpersonal Communication and Negotiation

Body Language 101. Lambert D. New York: Skyhorse Publishing; 2008.

Crucial Conversations: Tools for Talking When the Stakes Are High. Patterson K, Grenny J, McMillan R, Switzler A. New York: McGraw-Hill; 2002.

Flipping the Switch. Miller J. New York: G. P. Putnam's Sons; 2006.

Getting More: How to Negotiate to Achieve Your Goals in the Real World. Diamond S. New York: Crown; 2010.

Getting Things Done. Allen D. New York: Penguin Books; 2001.

Improving Business Communication Skills. Roebuck D. 4th ed. Upper Saddle River, NJ: Prentice Hall; 2005.

Interplay: The Process of Interpersonal Communication. Adler R, Rosenfeld L, Proctor II R. New York: Oxford University Press; 2010.

Teaching and Learning Communication Skills in Medicine. Kurtz S, Silverman J, Draper J. 2nd ed. London: Radcliffe Publishing; 2005.

Writing and Speaking for Excellence: A Guide for Physicians. St James D. Boston: Jones & Bartlett; 1996.

Writing, Speaking, and Communication Skills for Health Professionals. Barnard S, Health Care Communication Group. New Haven: Yale University Press; 2001.

Books on Presenting

The Naked Presenter: Delivering Powerful Presentations with or without Slides (Voices That Matter). Reynolds G. Berkeley: New Riders Press; 2010.

Presentation Zen: Simple Ideas on Presentation Design and Delivery. Reynolds G. Berkeley: New Riders Press; 2008.

Books on Writing

Engineering Your Writing Success: How Engineers Can Master Effective On-the-Job Communication Skills. Vincler J, Vincler N. Belmont, CA: Professional Publications; 1996.

Essentials of Writing Biomedical Research Papers. Zeiger M. 2nd ed. New York: McGraw-Hill Professional; 1999.

From Idea to Funded Project: Grant Proposals That Work. Belcher J, Jacobsen J. 4th ed. Phoenix: Oryx Press; 1992.

Get Funded!: A Practical Guide for Scholars Seeking Research Support from Business. Schumacher D. Thousand Oaks, CA: Sage Publications; 1991.

The "How To" Grants Manual. Bauer D. Lanham, MD: Rowman & Littlefield; 2009.

How to Write and Publish a Scientific Paper. Day R, Gastel B. 6th ed. Westport, CT: Greenwood; 2006.

On Writing Well. W. Zinsser. 7th ed. New York: Collins; 2006.

The Oxford Book of Modern Science Writing. Dawkins R. New York: Oxford University Press; 2009.

Proposals That Work: A Guide for Planning Dissertations and Grant Proposals. Locke L, Spirduso W, Silverman S. Thousand Oaks, CA: Sage Publications; 2007.

Publication Manual of the American Psychological Association. 6th ed. Washington, DC: American Psychological Society; 2009.

Scientific English. Day R. 2nd sub ed. Westport, CT: Greenwood; 1995.

Style Manuals and Guide Books

There are several good style guides available. See the list below for our favorites. Choose your manual to coincide with your audience and subject. Remember, the key is to be consistent throughout the document.

The AMA Manual of Style. 10th ed. Oxford: Oxford University Press; 2009.

The Chicago Manual of Style. 16th ed. Chicago: University of Chicago Press; 2010.

The Elements of Style. Strunk W Jr, White E B, Angell R. 4th ed. New York: Longman; 1999.

Grammar Girl: Quick and Dirty Tips for Better Writing. Fogarty M. New York: Holt Paperbacks; 2008.

Medical English Usage and Abusage. Schwager E. Phoenix: Oryx Press; 1990.

Online Information

Graphic Design Tips

Before & After. Available at http:www.bamagazine.com

Writing Tips for Communicating with the Lay Audience

The Center for Plain Language. Available at http://www.centerforplain language.org

News and Podcasts

National Public Radio. Available at http://www.npr.org

Writing Tips and Podcasts

Grammar Girl: Quick and Dirty Tips for Writing. Available at http://grammar.quickanddirtytips.com/

Presentation Tips

http://www.presentationzen.com/

Associations for Medical and Technical Writers

American Medical Writers Association
www.amwa.org

The AMWA provides support, classes, and networking for those who write or are employed in the field of biomedical communications.

Council of Science Editors
www.cse.org

CSE's mission is to serve editorial professionals in the sciences by creating supportive networks for career development, providing educational opportunities, and developing resources for identifying and implementing quality educational practices.

Society for Technical Communications
www.stc.org

The largest organization of its kind, the STC is dedicated to advancing the arts and sciences of technical communication. Its members include technical writers and editors, content developers, documentation specialists, technical illustrators, instructional designers, and web designers, developers, and translators—anyone whose work makes technical information available to those who need it.

Index

management, *see* serving
manners: business, 57; social, 143;
  telephone, 85
margins, 51
me; *myself*; *I*, 24–25
media, meeting with, 132–136;
  asking for tips, 133–134; checking the show in advance, 133;
  homework for, 133; notes for,
  134; organization policy for,
  132; podcasts, 136–137; preparation for, 133; print vs. broadcast, 134; requests for comments, 133; sound bites, 135;
  thank-you notes, 135–136;
  VODcasts, 136–137, 138; word
  choices, 134–135
meeting new contacts, 128–131;
  conciseness, 129; conversation
  skills, 129; conversation starters, 130–131; entering the
  room, 128; handshake, 128;
  humility, 129–130; listening,
  131; moving on, 131; nametag,
  128; "nice to see you," 129;
  open-ended questions, 130;
  professionalism, 131; smiling,
  128; social media, 130
meetings, 118–138; action plan,
  121; agenda, 120; arriving on
  time, 120; being prepared, 120,
  125; with colleagues, 119;
  encouraging participation in,
  121; excusing yourself from, 5;
  eye contact in, 122–123; hand-

outs, 123; how to run, 119–121;
  nonverbal signals, 127; note-
  taking in, 125; at other person's
  office, 6–7; in project management, 157; regular, 157; relevant
  contributions in, 126; respect
  for others in, 126–127; simple
  presentations in, 126; small-
  group, participating in, 125–
  127; small-group, presenting
  in, 122–125; specific points,
  126; staying on task, 120–121;
  staying on time, 120; terminology in, 126; thanks to participants, 121
Mehrabian, Albert, 78
Meng, Peter, 136
mentors, 142–143, 155, 167
micromanagement, 162
microphone, cordless, 105
mind mapping, 4
mini-quiz, 103
mirroring, 72–73
misplaced modifiers, 31–32
mistakes, correcting, 161
misused words, 22–26; *affect* or
  effect, 25; *between* or *among*, 26;
  ensure, *assure*, or *insure*, 23;
  fewer or *less*, 25; *i.e.* or *e.g.*, 23;
  in regards to, 26; *it's* or *its*, 22;
  me, *myself*, or *I*, 24–25
modal ("helping") verbs, 41
modifiers, misplaced or dangling,
  31–32
mouth, opening wider, 112

About the Authors

Stephanie Roberson Barnard and Deborah St James understand the nuances of communicating with well-educated, highly skilled people in science, technology, and medicine. As consultants for multiple high-tech companies, including Pfizer, Glaxo-SmithKline, AstraZeneca, Adobe, and Bayer, and universities, such as Harvard, Duke, Johns Hopkins, and Stanford, they have developed successful training programs to help professionals achieve their career goals. Like this book, their seminars are designed to maximize learning in the minimal amount of time by linking typical behavioral competencies with effective communication skills.

Since the early 1990s Barnard and St James have presented custom-designed writing, speaking, and communication seminars at hospitals and health maintenance organizations; medical, nursing, and pharmacy schools; pharmaceutical and biotech companies; and regional, national, and international science meetings. They have written patient education literature, edited manuscripts for journal publications, and developed marketing proposals and grant applications, as well as conducted hundreds of one-on-one tutorials.

Deborah St James (left) and Stephanie Roberson Barnard have collaborated since 1996. Both authors are "dog people," so it is fitting that Deborah's dog, Libby, decided to join them for this photo shoot and brainstorming session. Photo by Patrick Burns.